THE
GREAT PHYSICISTS

THE
GREAT PHYSICISTS

BY

IVOR B. HART

WITH 25 DIAGRAMS

Essay Index Reprint Series

Originally Published by
METHUEN & CO. LTD.

BOOKS FOR LIBRARIES PRESS
FREEPORT, NEW YORK

First Published 1927
Reprinted 1970

STANDARD BOOK NUMBER:
8369-1656-5

LIBRARY OF CONGRESS CATALOG CARD NUMBER:
71-117804

PRINTED IN THE UNITED STATES OF AMERICA

PREFACE

PHYSICS is a vast subject, and the story of its progress from early times is therefore lengthy. Yet it is well worthy of study. In this small work no intensive study is possible. Our task has rather been, therefore, to indicate its range in certain aspects, and to pay tribute to its great exponents through the ages by touching upon the lives and work of a few of its representatives. Sketchy as the book necessarily is, it is hoped that the reader will find reasonable continuity in the narrative from the days of Classical Antiquity onwards to modern times. No claim is made to originality, beyond perhaps that of treatment, and the author has drawn freely from many sources. If the reader will have been prompted to delve more deeply into the rich store of the scientific literature of the past after reading this book, its author will have adequately achieved his purpose.

Acknowledgements are due to Mr. E. D. Williams, M.A., for his careful and helpful reading of the manuscript, and to Dr. Charles Singer and Miss D. Turner, M.A., B.Sc., for the reading of the proofs, and to Mr. C. N. Heath, M.A., for assistance with the diagrams.

IVOR B. HART

Amersham
October, 1927

CONTENTS

LIST OF DIAGRAMS

THE GREAT PHYSICISTS

CHAPTER I

THE PHYSICISTS OF CLASSICAL ANTIQUITY

I. INTRODUCTION

THERE are many factors that contribute to the
building up of what is called civilization, and
one of these is undoubtedly Science. The quest of
science is the quest of truth ; and above all things the
seeking of truth has been, and must continue to be, the
most civilizing factor in the evolution of human society.
Our purpose in these pages is to review the main features
in the evolution of physical science by a consideration of
the life and work of some of the illustrious men who have,
through the ages, made their powerful and lasting contributions to knowledge.

Physics, however, has a specific meaning to-day such as
it could not possess in the days of antiquity. Knowledge
had not then reached the stage at which it could usefully
be divided up into specialized branches. All inquiries
into Nature and her mysteries were grouped under the
heading of philosophy, and of necessity the speculations
of the earlier philosophers were broad and all-embracing.
Nevertheless, inasmuch as these speculations became the
starting point of the investigations which have led to the
physics of to-day, no such survey as that upon which we
are now entering can fairly ignore them. Especially is
this true because the philosophers of antiquity included
some of the greatest minds in the history of civilization.

It will, however, be helpful as a preliminary to have before us a working definition of what is understood by Science, and although many alternatives have at one time or another been suggested, it will suffice for us to bear in mind that the object of science is to embrace in the smallest possible number of general truths the whole of the facts of nature. But an ability to generalize involves the handling of abstract ideas, and clearly there was a time when abstraction in thinking was unknown.

Man's first home was the tree and cave; his workshop was the surrounding jungle; his first tools were the nails, limbs and teeth with which nature had provided him. Yet one of the first steps in the dawn of reason came from the experience that taught man that there were other and often more effective weapons at hand in the boughs broken off from the trees around him, and in the stones lying at his feet. Thus the first lines of development were in the fashioning of better and sharper and stronger weapons.

What then was the next stage? There inevitably came a time when there was a general desire to refer back for explanations. It was a desire that, by the very nature of the human being, was bound to come, and it has been with us ever since. The early answers, moreover, were simple and comprehensive. They were based on superstition and the fear of what we now know to be natural phenomena on a grand scale—tempests, floods, thunder and lightning and so on.

There was always the latent desire also to associate with these phenomena the immediate action of powerful but invisible beings. As the poet has put it:

> Did raging storms o'er ocean's bosom sweep?
> 'Twas angry Neptune smote the troubled deep.
> Did clouds condensed emit electric fire?
> 'Twas Jove's wide-wasting instrument of ire.
> Did crops luxuriant fertile fields adorn?
> 'Twas Ceres decked the vales with wavy corn.
> Or Bacchus bade the high-embowering vine,
> Loaded with clusters, round the elm entwine:
> But if they perished by untimely blight,
> The Furies tainted the cold dews of night.

Thus we have it—the beginnings of the gods and the satisfying of the first dawn of the human intellect in the seeking after reasons. So gradually observation progressed and became orderly and ordered. Abstract reasoning had not yet arrived, but deduction based on practical needs was beginning to arrive. The older civilizations of the Chinese, the Indians, the Chaldeans and the Egyptians were building up astronomical data, schemes of number, and facts of practical geometry.

Nevertheless, we cannot say of these peoples that they were abstract thinkers. In the world of philosophy and science, their importance lies chiefly in that, from their large store of practical knowledge, they provided the great Greek nation with the stimulus for the first real efforts at abstraction and philosophical speculation.

We may regard the Greeks, then, as having initiated true science in the modern sense of the term. Keen of intellect, temperamentally of an inquiring nature, they were naturally well fitted for the task of inaugurating the era of science for the world. Certain geographical and economic advantages operated in support of this. With its great seaboard and its famous Ionian Islands, it was but natural that the Greeks should be both seamen and traders, and indeed the ships of such cities as Miletus were to be seen in every port in the Mediterranean. The Ionians, in particular, showed a love of adventure that prompted them to leave their homes and found colonies. As a consequence they were able to come into contact with other minds and other peoples, to absorb from them such learning as they possessed and to apply to this knowledge their own fertility in speculation and generalization.

II. THALES OF MILETUS

This was not strictly true in the case of Egypt, which, for some time prior to the seventh century B.C., had adopted a policy of isolation from the rest of the world. Egypt had been suffering many vicissitudes and internal dissensions, and was beginning to recover at about 700 B.C., and it was in about 670 B.C. that the Greeks were offered friendly access to her shores. It is not, therefore, without significance that the great days of Ionian Physics date from the sixth century B.C., and begin with Thales of Miletus, a philosopher who was born about 636 B.C. and who died about 546 B.C. There is no doubt as to his genius. A man of civil affairs, an engineer, a philosopher and scientist, he seemed to excel in everything. As a philosopher his work lay along three main paths —geometry, astronomy and physics. It is with the third of these that we are here concerned. To Thales belongs the credit of the first observations on record in the subject of electricity. He observed that when amber is rubbed, it attracts light substances like pieces of paper, and it is from the Greek word for amber, *elektron*, that the modern term electricity is derived. He also knew of the magnetic property of the mineral called 'lodestone' or magnetite, though to both these phenomena he attributed a certain quality of 'soul' or 'animation.' However, the chief claim of Thales to our attention is his famous speculation as to the common origin of all things from one material. 'All things,' he said, 'are produced from water.' It is of little consequence that Thales was hopelessly wrong. What does matter enormously is that here we have for the first time a general speculation of the first magnitude—the reference back from effect to cause of all things and all phenomena to a single primary basis. It was in fact the first great generalization as to the constitution of matter in the history of science, and it was the starting point of a whole series of similar speculations which formed so characteristic a feature of Greek philosophy.

For example, Thales' own contemporary and disciple, Anaximander, and his successor Anaximenes, both offered variations on this subject as further contributions. The urge to find some primordial substance had certainly come to stay, although the particular substance or substances might and did differ as between one philosopher and another. Thus Anaximander held the primordial substance to be something more subtle than water, though not so thin as air, whilst his successor, Anaximenes, went the one step further and taught that ' all comes from air, and into air returns.'

III. PYTHAGORAS OF SAMOS

Pythagoras, another great Greek philosopher, was born in the Ionian Island of Samos about 580 B.C. He was not at first a great success. Indeed there is a story that at Samos he only had one disciple who at first only consented to listen to Pythagoras on being promised a small coin every day. But Pythagoras, seeing after a time how interested this pupil became, in his turn refused to lecture, whereat the disciple intimated that he would in his turn pay Pythagoras a similar sum to go on lecturing, and so our philosopher got his money back. He eventually left Asia Minor and settled in Croton, a Dorian colony in Southern Italy. Here he established a school of philosophy and his followers became known as the Pythagoreans. They were a mystical brotherhood, and their studies of mathematics and science were apparently incidental to their moral and religious doctrines.

There is some difficulty as to what teachings may legitimately be attributed to Pythagoras himself on account of the practice of his school of committing nothing to writing. All its learning was passed on by word of mouth, so that it is difficult to say how much of its doctrine may be attributed to Pythagoras himself, and how much to his disciples. Like their predecessors, they sought for first principles, but they found them in the less material direction of order, harmony and proportion. Through this agency they gave a great stimulus to the study of number. In this they appeared to find the essence of all things. Ten was a perfect number, because it was the sum of $1 + 2 + 3 + 4$. Three was also sacred to the Pythagoreans—it was the ' number of the universe,' because everything had its beginning, its middle and its end.

We must not omit to mention, also, the inauguration by Pythagoras of the scientific study of sound. The story goes that while passing a blacksmith's shop he was attracted by the musical notes emitted by the anvil on being struck with the hammer, and subsequent experiments on stretched strings of various lengths on what to-day we speak of as the monochord, established the relations to one another of what are now known as the musical intervals. Pythagoras found by experiment that the necessary length ratio to produce an octave was 2 : 1; to produce the interval known as a fifth was 3 : 2; and to produce a fourth was 4 : 3. So he got back to his magical series of 1, 2, 3, 4. Small wonder, then, that he found himself back to his great doctrine of harmony and proportion, and that its powerful appeal should have led him to apply this to his famous conception of the harmony of the spheres—the picturesque scheme of a central solar fire at the centre of the universe, with the planets revolving round them with varying speeds that created celestial notes in harmonic ratios excelling in sweetness all earthly music—so sweet, indeed, as to be inaudible to mankind.

We find a reference to this by Pope in his *Essay on Man* :—

> If nature thundered in his opening ears
> And stunned him with the music of the spheres,
> How would he wish that heaven had left him still
> The whispering zephyr and the parting rill !

Like his predecessors, Pythagoras speculated upon the structure of matter. His contribution, however, was a very important one, and was the starting point of a theory that developed and persisted right through the Middle Ages. This was the theory that all matter was made up of the four elements of earth, air, fire and water. The theory was further developed by Empedocles of Agrigentum, a disciple of Pythagoras, who taught that these four primary elements are imperishable and unchangeable in quality, and he explained that they were brought into various combinations by the agency of two divine forces—one of attraction, love, and the other of repulsion, discord. In this way he supposed the elements to combine in varying proportions to produce all the natural phenomena with which the world was endowed.

Whatever the merits or demerits of this theory, it will be seen at any rate that the Greek philosophers were now immersed in a line of inquiry which is to-day occupying a great portion of the field of research of the modern physicist—the constitution of matter. In fact in the hands of those who followed Empedocles we are even more startlingly reminded of the trend of modern research. Leucippus of Abdera, for example, breaking clean away from the Pythagorean teachings, declined to accept the four-element theory, and substituted for it the notion that all things are formed from particles so small, that they could individually be neither seen nor felt. As he conceived them, however, they were of various forms and sizes, and it was their difference in arrangement and movement that produced the various complexities of matter. This atomic theory later received further stimulus at the hands of Leucippus' pupil, Democritus of Abdera, who specifically enunciated the principle of the indestructibility of matter. ' Nothing can be made from nothing,' he said, and all change consists only in the composition and the separation of particles. The world is a scheme of atoms and empty spaces. Unfortunately the views of this atomist school soon became submerged in the teachings of the giant Greek personalities who followed, but it surely remains one of the wonders of the history of science that now, after more than two thousand years have elapsed, we have virtually returned to these views, and have accepted a scheme of chemistry based almost entirely upon them.

IV. THE PERIPATETICS

We now come to a period in Greek science which, chiefly by virtue of the profound influence it exerted in dominating the whole intellectual outlook of the Middle

Ages for very many centuries, is easily the most important in the science of Classical Antiquity—the era of Plato and Aristotle.

The centre of Greek learning had by now shifted from Ionia and Southern Italy to Athens, and it was here that the immortal Socrates thought and taught ; it was here too, that his pupil Plato (427–347 B.C.) founded his famous Academy, in a pleasant grove with shady walks and seats, and with the famous inscription over the doors that read ' Let no one presume to enter here who has no taste for Geometry and the Mathematics.'

Plato's works, written mainly in the form of dialogues, constitute one of the main legacies which the sons of Greece left to civilization, and from the point of view of Science perhaps the most important of these was the "Timaeus."

Plato also gave to the world a number of earnest students of philosophy whom he had taught, and who were able in their turn to make their own contributions to knowledge. The most important of these was the philosopher who, beginning as a student in Plato's Academy, became in his turn an influence of the highest significance in the scientific history that followed.

Aristotle was born in the city of Stagira, in Macedonia, in the year 385 B.C. His father was Nicomachus, a physician. He entered Plato's Academy as a student at the age of seventeen, where it is related his thirst for learning was so great that he resorted to various devices to reduce his hours of sleep.

Such zeal was bound to bring its reward. In due course Aristotle left the Academy. Later he founded a school of his own. In consequence of his habit of discoursing to his pupils whilst walking up and down a fine avenue leading from a covered portico or ' peripatos,' he has been spoken of as the Peripatician and his followers are sometimes referred to as the Peripatetic Philosophers.

Let us first of all consider Aristotle's speculations as to the constitution of matter. Here his work was further to elaborate upon the four-element theory so that these came to be spoken of afterwards as the Aristotelian elements. In addition to the four elements there were four primary qualities : heat, cold, moisture and dryness ; and each element was considered to be compounded in its turn of two of these primary qualities in what he called ' binary combination.' Thus earth was compounded of cold and dry ; water was compounded of cold and moist ; air was compounded of hot and moist ; fire was compounded of hot and dry.

Further, whereas Aristotle held that all terrestrial matter was built up of earth, air, fire and water, for celestial matter he introduced the idea of a fifth element —the *quinta essentia*, the ether more subtle and divine than the other elements. This gives rise to the word ' quintessence ' of to-day. It carried with it for Aristotle this important distinction : that whereas terrestrial matter implied a variety of possible combinations and disintegrations within the limits of the four elements, there could be no change in celestial matter. The skies, being perfect, were immutable, and there could therefore be no break in their uniformity.

Aristotle's views on the subject of motion are of extreme interest.

He begins by stating that as lines are measurable in one direction only, and planes in two, so bodies are measurable in three directions. Body is therefore made complete by three magnitudes, and three is therefore the number of perfection.

As matter, then, is made complete by three, the same must also be true for motion. There are therefore three simple motions—(*a*) motion towards the centre, (*b*) motion from the centre, and (*c*) motion round the centre. The first two are rectilinear motions, and are natural to earthly substances, whilst the circular motion appertains to the heavenly bodies.

So far as motion towards and from the centre are concerned, we have here a sort of theory of gravity. Plato had already attempted such a theory, arguing that a downward motion was really a motion towards the centre of the earth, and that there existed a tendency for all bodies to be attracted towards larger masses of the same material. According to Plato, for example, a lump of rock or stone is attracted down to the earth, whilst a 'vapour' rises by attraction to the larger masses of vapour above.

Aristotle took a different view. He argued that there were two classes of bodies—heavy bodies, which have a natural tendency to move down towards the centre of the earth, and light bodies, whose natural tendency is to move up from the centre of the earth. Further, he held that there were degrees of heaviness, the heavier bodies tending to move further down than the less heavy bodies, so that, e.g., earth, being heavier than water (both in the 'heavy' class of matter), we expect to find the water above the earth, as indeed we do. On the other hand, both 'fire' and 'air' belong to the 'light' group, and since fire is lighter than air, both have a tendency to go up, but the fire more so than the air, and so the air is above the water, and the fire nearest the celestial regions.

One of his deductions as to falling bodies was that a heavier body would fall faster than a lighter one. This was destined to prove a serious obstacle to progress in mechanics for very many centuries, and it is somewhat astonishing that, with his vigorous mind, it should not have occurred to him to put such a simple problem to the test of experiment.

Aristotle died at the age of sixty-two years at Chalcis, leaving behind him a record of study and wisdom that has earned for him the lasting homage of mankind.

V. ARCHIMEDES (c. 287–212 B.C.)

We turn finally to one who was perhaps the greatest exponent of physical science in classical antiquity, Archimedes. He was born at Syracuse, in Sicily, about 287 B.C., and received his training and education at the famous school of Alexandria founded by the second of the long line of Ptolemys who ruled in Egypt. The Ptolemys were great lovers of learning. They invited to their court a large number of scholars and philosophers, and equipped for them lecture rooms, libraries, gardens and observatories. It is not surprising, therefore, that Alexandria became the chief centre of culture and science, and that the Alexandrian School should have given to the world a succession of great thinkers that included Euclid the Geometer, Aristarchus the Astronomer, Eratosthenes the Geographer, and in its later days Hipparchus, Claudius Ptolemy, and Apollonius of Perga.

Archimedes belongs to the earlier period, and indeed it is not improbable that Euclid was one of his teachers. However, he did not stay in Alexandria, but returned to Syracuse, and here he spent the remainder of his life in meditation and study. In one notable respect he stands supreme among Greek scientists. He was definitely an experimenter. Greek philosophy was characterized chiefly by its speculativeness, and there are very few instances, except possibly in the field of astronomy, of the experimental method among their achievements. Archimedes was a brilliant exception. He fully understood the test of experiment, and indeed as an experimental physicist he almost stood alone for very many centuries.

The world owes very much to Archimedes. Like Euclid, he was a great mathematician, and although he is more generally known for his work in mechanics, he himself was very proud of his mathematical discoveries. One of these was his elucidation of the relation between the areas and volumes of a sphere and its circumscribing cylinder, and he was so proud of this that he expressed a desire for these figures to be recorded on his tomb.

In physics the greatest achievements of Archimedes were in the particular branch known as mechanics. The doctrine of equilibrium was treated by him in a masterly manner. His starting point was the principle of the lever, the demonstration of which was made to rest ultimately on the truth that equal bodies at the ends of the equal arms of a rod, supported at its midpoint, will balance each other. Proceeding from this, he

2

pursues his proof to the conclusion that bodies will be in equilibrium when their distances from the fulcrum (i.e., the point of support) are inversely as their weight. Hence, given a sufficiently long lever, any weight, however big, may be suitably moved. ' Give me,' said Archimedes to his King, ' where to stand, and I will move the earth.'

From this he passed naturally to a discussion on centres of gravity, establishing several propositions relating to them, and working out the positions of the centre of gravity for variously shaped bodies.

Archimedes, too, was virtually the founder of the branch of physics known as hydrostatics, i.e., the study of the laws of fluids at rest. The story of his discovery of the famous Principle of Archimedes—namely, that a solid body, when immersed in a liquid, appears to lose a portion of its weight equal to the weight of liquid it displaces—has indeed, in one form or another, become a classic. It appears that the principle occurred to him as a result of his observations of the water rising in his bath on his immersion therein, and that he was so excited at the discovery that he ran out without stopping to clothe himself, exclaiming, ' Eureka ! ' (I have discovered it). The story usually linked up with this relates how a golden crown, ordered by King Hiero of Sicily, was suspected of adulteration in its manufacture, and Archimedes was invited to find a means of proving or disproving this. After the reputed incident of the bath, he was able to base his solution of the difficulty on the fact that when two masses of equal weight, but of different density or specific gravity, are successively immersed in water, the less dense, being the larger, will displace a larger body of water. Hence if the adulterated crown contained some metal lighter than gold, it would displace a greater quantity of water than a crown of pure gold of the same weight.

Archimedes excelled in the design of mechanical contrivances both for peace and war. His famous ' screw,' virtually a pump and used for raising water from a lower level to a higher one, is applied in practice to this very day. In effect it is a pipe twisted in the form of a corkscrew and is held in an inclined position with one end immersed in the water source below. It is rotated on its axis by means of a handle at the top, and the water is thus brought up with the successive turns of the screw and ultimately runs out at the top.

Archimedes' death occurred in 212 B.C. under tragic circumstances at the taking of Syracuse by Marcellus, during the second Punic War. Marcellus, with commendable reverence for learning, had ordered that the philosopher should be spared. Unfortunately, he was discovered by a soldier who did not recognize him, and, according to one account, Archimedes was so engrossed in a mathematical problem the diagram for which he had figured in the sands that he did not hear the angry calls of the soldier, who thereupon slew him.

CHAPTER II

ISLAMIC PHYSICS—ALHAZEN AND ROGER BACON

I. THE SECOND SCHOOL OF ALEXANDRIA

AS a result of the ascendancy of the Roman Empire, Egypt, the home of Greek culture, became a Roman province. Now an occurrence of such magnitude was bound to have an effect. Greek philosophy under the stimulus of freedom was one thing, but under the subjection of foreign rule it became a totally different thing. It began to lack initiative and originality, and to show signs of decline. Unfortunately, too, a careful study of their writings shows that great as they were in war, in polite literature, in government, and in law, the Romans were singularly lacking in a disposition towards the study of the sciences.

In view, then, of these great changes in the circumstances of the Greeks, it is not surprising that the high fame of the great school of Alexandria began to suffer a serious decline. Luckily, before this decline could reduce it to anything like extinction, there arose, about the year A.D. 140 a family of Roman rulers, the philosophical Antonines, under whose influence there began a distinct revival of the study of science. This revival gained considerable impetus under the rule in particular of the famous Marcus Aurelius Antoninus, and as a result a fresh series of brilliant men of science gave a new lease of activity to what came to be known as the Second School of Alexandria.

It was to this period that Claudius Ptolemæus, or

Ptolemy, as he came to be commonly called, belonged.
He was a native of Egypt and taught at Alexandria
during the second half of the second century A.D.   He
is known chiefly as the author of the *Syntaxis*, (or
Algamest), the great astronomical work that supplied
a geocentric theory of the universe which persisted to
the time of Copernicus in the sixteenth century.   So far
as physics is concerned, however, he also contributed
important researches in optics.   These will be discussed
later in this chapter.   Apart from Ptolemy's optics, the
main work of the Second School of Alexandria was in
astronomy and mathematics.

II.   THE DARK AGES

In due course the Roman Empire in its turn began to
suffer decay and disintegration, with results that threatened a very black future for learning in Europe.   The
division of the Empire under the sons of Theodosius,
and the continual wars with surrounding enemies, produced an inevitable effect upon culture and gradually
brought about the extinction of the School of Alexandria
as a centre of learning.   The buildings were slowly
suffering destruction, and the books were meeting the
same fate.   Alexandrian conditions were only sympto-
matic of what was happening elsewhere.   Cut off from the
incentive of Greek learning, the whole of Europe received
an educational set-back from which it was not to recover
for some hundreds of years.   And so the decline and
dissolution of the Roman Empire brings us to that
period in the history of learning of which we speak as
the Dark Ages.   In the general strife and unrest of those
days, we have, so far as Europe is concerned, a long and
unhappy interval of darkness and ignorance and neglect.
There were no original writers, and very few readers.
Such little learning as existed was confined to the clerics
such as Isidore of Seville (A.D. 560–636), the Venerable
Bede (673–735), Alcuin (735–804), Raban (786–856) and
others, and these confined their efforts to the writing of

crude Encyclopædias embodying poor Latin versions of
such works of Aristotle, Plato and Pliny as were available.

III. THE ADVENT OF ISLAM

Just as learning was almost at the point of extinction
in Europe a hope for its revival appeared from a totally
fresh quarter. The new channel of knowledge was
Islam. In the year A.D. 622 Muhammad, whose great
miracle was the inspiration of the warring tribes of
Arabia with common ideals, sent them forth to conquer
half the then-known world, and to found a mighty empire
whose centre was first at Mecca and later at Medina.
They took possession of Alexandria in 640 A.D. and
promptly burnt what was left of the famous old library ;
they overran all the north of Africa up to the straits of
Gibraltar, and they even reached parts of Spain and
France. There followed a period of rule under the line
of Umayyad Caliphs, whose immense Empire stretched
from Spain to Samarkand, and whose court at Damascus
soon began to show a luxury and wealth hitherto un-
dreamed of by the Arab.

What was the influence of all this on learning ? There
is a popular idea that the Saracens were lovers of learning. This is not strictly correct. What happened was
that amongst the subject peoples under the Saracen
rule were a number of Arabic-speaking races who were
students by nature and philosophers by inclination.
Such were the Nestorians, the Jews, the Persians, the
Syrians and the Moors. These peoples seized on such
treasures of Greek literature and learning as had been
salved from spoliation, and at once proceeded to translate
them into Arabic. Their activities were at first regarded
with suspicion by their rulers, but after a time there
came to the throne a succession of Caliphs who patronized
and encouraged learning. Amongst these were the
Caliphs El-Mansour and Haroun-el-Raschid in the
seventh century A.D., and the Caliph El-Mamoun in the

eighth and early ninth, and it was this latter person who in A.D. 827 ordered to be translated into Arabic Ptolemy's famous *Syntaxis*, under the title of *Almagest*. There were at various times centres of learning at Meragha in Persia, Baghdad in Mesopotamia, Samarkand, and at Cordova and Toledo in Spain.

IV. ARABIAN PHYSICS—ALHAZEN AND HIS PRE-DECESSORS

Generally speaking, Arabian science was very active in alchemy and medicine, but so far as physics is concerned it was barren of any advance on the science of the Ancients except in one field—that of optics ; and the great exponent of this study was Alhazen. He was a native of the city of Basra, in Mesopotamia, and was born in A.D. 965. Very little is known of his life except that, having boasted that he could construct a machine for regulating the inundations of the Nile, he was ordered to Egypt by the Caliph Hakem Bi-Amrillah and given the opportunity to make good his boasts. He failed to do so, and avoided punishment by fleeing to Cairo and feigning madness until the Caliph's death in 1021. Alhazen himself died in 1038. He was a most diligent student, and his work in optics was outstanding. As a preliminary to the consideration of this it will be advisable to review in brief the main lines along which previous researches in this subject had been pursued.

The first treatise in optics was the product of the philosophers of the Alexandrian School, a fact fitting naturally with their well-known activities in geometry, since the regularity with which the rays of light take rectilinear paths would call for geometrical treatment. Euclid had stated that light travels in straight lines, and he investigated the principles upon which the magnitudes of objects and their images are judged, though his treatment was very defective. Practical optics, on the other hand, was considerably in advance of theory, and among the ancients had certainly reached the stage of construction of metallic mirrors, both plane and spherical. Various references occur to the use of glasses in spherical form for burning by concentrating the rays of the sun on to a point.

The earliest is by Aristophanes in the *Clouds* (*c.* 424 B.C.) in the following dialogue :— [1]

Strepsiades. I have found a very clever method of getting rid of my suit, so that you yourself would acknowledge it.

Socrates. Of what description ?

Strepsiades. Have you ever seen this stone in the chemists' shops, the beautiful and transparent one, from which they kindle fire ?

Socrates. Do you mean the burning-glass ?

Strepsiades. I do. Come, what would you say, pray, if I were to take this, when the clerk was entering the suit, and were to stand at a distance, in the direction of the sun, thus, and melt out the letters of my suit ?

Socrates. Cleverly done, by the Graces.

Strepsiades. Oh ! How delighted I am, that a suit of five talents has been cancelled.

Pliny mentions the power of a globe of glass, filled with water, to produce heat, and expresses surprise at the fact that the water itself should throughout remain quite cold. Seneca also gives a number of optical references in his *Natural Questions,* among them a comment on the magnification of vision through glasses. He describes, for example, an angular glass rod, evidently of prismatic form, which produces colours like that of a rainbow. This he attributes to the irregular formation of the image, and adds that such a glass, properly made, will give as many images of the sun as it has angles ; which is nothing more than the multiplication of an object by a glass cut with a number of faces.

A much more serious contribution to optics was offered by Cleomedes early in the first century of the Christian Era. To the previous work on reflection he added a study of refraction, i.e., the bending of light on its passage from one medium to another. He des-

[1] Taken from C. Singer—*Studies in the History and Method of Science,* Vol. 2, p. 386.

cribed the phenomenon of the bent appearance of a
rod partially immersed in water. From this he passed
to the experiment of pouring water on to a coin at first
invisible at the bottom of a basin, until the water
reaches a depth at which the coin just becomes visible
by refraction. Cleomedes was able from this to deduce
that the sun was visible by atmospheric refraction even
when below the horizon.

The greatest advance in optics among the ancients
was, however, made by the famous Ptolemy. His
contribution was also mainly on the subject of refraction. Its importance to Ptolemy was naturally
due to his activities as an astronomer, since by refraction through the atmosphere, as the light passes
through to an observer, the stars are seen deviated
from their true positions in the skies. Ptolemy examined with great care and accuracy the angles of refraction corresponding to all angles of incidence from 0° to
80° in the case of a ray of light passing from air to a
glass or water medium, and his tables of correctons for
the altitudes of the stars were extremely accurate having
regard to the circumstances. His result approximated
to those obtained from the law of sines as we know
them to-day, though of course Ptolemy was himself
unaware of the Sine Law. He also distinguished what
we now speak of as the virtual focus of a convex lens
—i.e., the point at which the diverging reflected rays
would meet, if produced backwards behind the lens.

V. ALHAZEN'S OPTICS

No further advance is to be recorded until Alhazen's
time. This philosopher wrote very profusely on the
subject. His writings in his famous book *Thesaurus
Opticæ* were both prolix and without method, but they
nevertheless display an originality and skill generally
superior to Ptolemy, to whose works he was greatly
indebted, except on the subject of atmospheric refraction as applied to astronomy. Especially refined were

his applications of geometrical methods to the elucidation of optical effects in curved mirrors. For example, he solved the problem of finding the point in a convex mirror at which a ray coming from one point shall be reflected to another given point. He offered the suggestion that the sun and moon on the horizon appear larger than in the zenith owing to the influence on our judgment by comparison with terrestrial objects. In his treatment of refraction he gave an explanation of the cause of twilight, and he considerably improved on Ptolemy's apparatus for measuring the angles of refraction in different media. His methods were in fact strongly reminiscent of the laboratory methods of to-day.

The *Thesaurus Opticæ* consists of seven books. The first discusses the nature of light and colour and explains the anatomy of vision. The second deals with the functions of sight and with the physiology of perception. The third discusses optical illusions. The next three deal with reflection, and the last with refraction. He discusses all kinds of mirrors, plane, spherical, cylindrical and conical, and includes both convex and concave types.

The fourth book is of special interest, detailing as it does the construction of apparatus for the verification of the law of reflection that the angle of incidence is equal to the angle of reflection. The extreme care shown in instrumental construction is wholly admirable. The orifices through which the rays were admitted were half a barleycorn (one-sixth of an inch) in diameter, but for special accuracy this orifice was closed up with a wax plug the centre of which was pierced by a needle point.

Contrary to the views current until his day, Alhazen held that vision resulted from rays coming from the object to the eye, and were not a result of emanations from the eye to the object. Alhazen, too, was considerably in advance of previous writers on the subject of the structure of the eye, placing the lens (the humor crystallinus) in the centre of the optic globe, and regarding it as the centre of conversion of the external light-rays into the sensation of sight.

Alhazen's works were translated into Latin by an unknown writer, and this Latin translation was made the basis of a detailed study about 1260 by Witelo, a Polish writer, whose book greatly influenced the famous Roger Bacon shortly after. Witelo's work was in fact little more than an unacknowledged revised edition of Alhazen. Incidentally his tables of refraction proved to be little more than a repetition of those of Ptolemy.

VI. THE RECOVERY OF SCIENCE TO CHRISTIAN EUROPE

Luckily for Europe, there were not wanting men who were curious about the new Arabic learning. Gradually, the news of this learning began to filter through to Europe as a result of the experiences of such men as Sabbatai ben Abraham (or, as he was more commonly known, Donnolo, a Jew of Otranto), who practised medicine at Rossano in Southern Italy. He was at one time during the first half of the tenth century a prisoner in the hands of the Saracens at Baghdad. A native of the town taught him Arabic and enabled him to study the new learning at first hand, and this he was able to pass on to the Western world on his release. Constantine the African was another early example. He was a native of Carthage, and of course spoke and wrote Arabic. He reached Italy about A.D. 1060, became a monk at Montecassino, and freely translated into bad Latin many current Arabic works in medicine and science. Yet another was Gerbert, a tenth-century monk of Aurillac, who afterwards became Pope Sylvester II. He spent some years at Barcelona, mastered Arabic, studied Arabian works of science and then returned to offer the fruits of his labours to the Western world.

All these and others of similar type helped in that slow infiltration of knowledge which was to bring about a new dawn for science in Europe. But what was the nature of the learning which was thus beginning to penetrate into Europe ? Remember that the ultimate sources were the old Greek books. These had been more or less indifferently translated into Arabic, perhaps through the intermediary of Syriac, and now these Arabic versions were in their turn being rendered into Latin. The chief centre for this was at Toledo in Spain, because, being further north than Cordova, it had remained largely Christian, though the general medium of speech was a sort of Arabic vernacular. It is scarcely surprising that after so many different translations, the original Greek versions were very much distorted, and as a consequence much passed as the teachings of the ancients which in fact those philosophers never intended. However, garbled as they were, the teachings of Aristotle, Plato, Ptolemy and others were once again rendered available to Europe. The Dark Age was passing, and the dawn of a new era of learning was begun. Universities began to spring up in the different countries. Students were attracted in increasing numbers, and by the thirteenth century it could be said that Arabian science had been completely transmitted to Western Europe. However, with all this increase in scientific interest, there was clearly something lacking. Slowly it began to be recognized that all these mis-translations with which the universities were being flooded were unsatisfactory and that the only true road to the wisdom of the ancients was the study of the Greek language. Foremost among those who pushed this plea were the Franciscan monks of the thirteenth century, and of these the famous Roger Bacon, with his keen intellect and powerful vision, towered head and shoulders above his fellows in learning, in influence, and in experiment. We hear much of the work of Roger Bacon, but probably it is not too much to say that his greatest work was his insistence on a proper study of Greek and Arabic with a view to enabling the student to go direct to the ancients for their wisdom.

VII. ROGER BACON

Roger Bacon was born in 1214, probably at Ilchester in Somerset. He appears to have come from a wealthy family, and he began his studies at home in 1227. Later he proceeded to the University of Oxford. Here he came under the influence of Edmund Rich, then lecturing on Aristotle, and Robert Grosseteste. This latter scholar was one of the thirteenth-century pioneers of the study of Greek, and under his influence Roger Bacon himself studied Greek, and afterwards entered the Franciscan Order. From Oxford Bacon proceeded to the University of Paris, at that time the most active of the European centres of learning, and here he devoted himself to a detailed study of science, his experimental methods earning for him renown as the 'Admirable Doctor.' He was, however, much too progressive in his views to suit the ecclesiastical authorities, and some years after his return to Oxford in 1250, he fell under the suspicions of the General of his Order, and he was banished to the confinement of a house in Paris. Here he stayed from 1257 to 1267. Towards the end of this term he wrote his three large treatises, the *Opus Maius*, the *Opus Minus* and the *Opus Tertium* on the invitation of Pope Clement IV. It was possibly owing to this Papal influence that he was allowed to return to Oxford in 1268. Here in 1271 he produced the first part of his *Compendium Studii Philosophiæ*. Once again, however, he incurred the displeasure of his order, and for fourteen years, from 1278 to 1292, he was again confined in Paris. He spent the last two years of his life in Oxford, and died in 1294 at the age of eighty. He was buried in the Church of the Franciscans.

Our debt to Roger Bacon is not so much for the mere details of the contributions he made towards knowledge as for his bold appeal to experiment and to the observation of nature. Nevertheless, it is impossible to withhold our admiration from him also as a wonderful visionary of the future. It is astonishing to think that the following famous passage could have been written, as it certainly was by Bacon, nearly seven hundred years ago :

' First, by the figurations of art, there may be made instruments of navigation without men to rowe them, as great ships to brooke the sea, only with one man to steere them, and they shall sayle far more swiftly than if they were full of men ; also chariots that shall move with unspeakable force, without any living creature to stirre them. Likewise, an instrument may be made to fly withall if one sit in the midst of the instrument, and doe turne an engine, by which the wings, being artificially composed may beat the ayre after the manner of a flying bird. . . .

' But physicall figurations are farre more strange : for by that may be framed perspects and looking-glasses, that one thing shall appeare to be many, as one man shall appeare to be a whole army, and one sunne or moone shall seem divers. Also perspects may be so framed, that things farre off shall seem most nigh unto us.'

Bacon's scientific activities covered a very wide range, and included mathematics, astronomy, geography, alchemy, astrology and physics. It is with this last that we are here concerned, and in conformity with the tendencies of the times, his main work was in the field of optics. Bacon, however, developed some remarkable views on the subject of the propagation of force that must first claim our attention. These are set forth in both the fourth part of the *Opus Maius* and in his *De Multiplicatione Specierum*. These views link up with his work on optics because, for Bacon, the radiation of light was a type of all radiant forces. He conceives an emanation of force to be continually proceeding from every bodily object in all directions. The first result of this emanation he calls the ' species,' or image, or impression. This is a doctrine frequently met with in varying forms among the writings of Lucretius, and other ancients. Bacon, however, was less crude and far more specific. For example, he rejects the notion that the species is something emitted from the acting body, since if it were so, it would be weakened and ultimately destroyed, which does not in fact happen. Nor does the body function in the rôle of a transmitter of the species. What happens is that the agent stimulates the potential activity of the matter acted on. The subject is excellently summarized by Bridges [1] thus :

'The agent acts on the first part of the body of the patient, and stimulates its latent energy to the generation of the species. That part thus transmitted acts on the part next succeeding; and so the action proceeds. While the agent acts on the patient the patient reacts on the agent. . . . There is in this way an interchange of force between all parts of the universe.'

'The ray, or species, is of corporeal nature; but this corporeal nature is not distinct from that of the medium ; it is generated from the substance of the medium, and is continually re-formed out of successive portions of the medium occurring in the line along which the force is propagated. If wind is driving the air transversely to the line of force, this in no way affects this line. The species is formed and re-formed from particles of the medium presented in the line of propagation, and from no others.'

'Finally, the propagation of rays occupies time, though its velocity is such that the time occupied in passing through so vast a space as the diameter of the universe is imperceptible to sense.'

Here, surely, we have a conception that has much in common with the undulatory theory, in that the 'species' is a motion or change in successive portions of the medium, propagated in straight lines, and deflected in direction when the medium changes. The time factor in transmission is especially significant. So, too, is Bacon's discussion of opaque bodies.

'No substance,' he says, 'is so dense as altogether to prevent rays from passing. Matter is common to all things, and thus there is no substance on which the action involved in the passage of a ray may not produce a change. Thus it is that rays of heat and sound penetrate through the walls of a vessel of gold or brass. It is said by Boethius that a lynx's eye will pierce through thick walls. In this case the wall would be permeable to visual rays. In any case there are many dense bodies which altogether interfere with the visual and other senses of man, so that rays cannot pass with such energy as to produce an effect on human sense, and yet nevertheless rays do really pass through without our being aware of it."

Obviously these views were intimately related to Bacon's work on optics generally. This subject absorbed

[1] Bridges : *Life and Work of Roger Bacon*, 1914, p. 97.

his attention for some ten years of his life, and fills a fifth part of his *Opus Maius*. Greatly indebted to Alhazen, he nevertheless shows great superiority over his Arabian precursor, and shows proof of a steady appreciation of the practical possibilities involved in the phenomena of reflection and refraction. He was certainly quite clear about the simple microscope. Thus he writes :

' If the letters of a book, or any minute objects be viewed through a lesser segment of a sphere of glass, or crystal, whose plane base is laid upon them, they will appear far better and larger. Because by the fifth canon about a spherical medium, if its convexity is towards the eye, and the object is placed below it, and between the convexity and its centre, all things

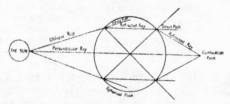

Fɪɢ. 1.—Roger Bacon's Diagram to illustrate his theory of the Burning-glass.

concur to magnify it. For the angle under which it is seen is greater and its image is also greater, and nearer the eye than the object itself, because the object is between the centre and the eye ; and therefore this instrument is useful to old men, and to those that have weak eyes ; for they may see the smallest letters sufficiently magnified.'

Bacon's ideas on refraction are well seen in his explanation of the action of the burning glass, which he illustrates by means of a diagram (Fig. 1).

' If any one hold a crystal ball or round flask filled with water in the strong rays of the sun, standing by a window in face of the rays, he will find a point in the air between himself and the flask at which point, if any combustible substance is placed, it will catch fire and burn, which would be impossible unless we suppose a double refraction. . . . Now, since an infinite number of rays are given off from the same point of the sun, and one only falls perpendicularly on the flask, all the others are refracted and meet at one point on the perpendicular ray which is given off along with them from the sun, and this point is the point of combustion. On it are collected an infinite number of rays, and the concentration of light causes combustion. But this concentration would not take place except by double refraction, as shown in the diagram.'

Although, too, he was aware of the possibilities of lens combinations, it cannot be said that he invented the telescope. He certainly introduced the possibility of distant vision by instrumental means, but it had to be left to a later generation to find a solution.

In an era devoid both of scientific personalities and scientific appreciation, Roger Bacon stands out as a vivid figure. Ahead of his contemporaries in outlook, it was inevitable that his great example should find little response. But it remains to his glory that in an age of intellectual slavery to tradition he upheld the claim to responsible criticism the insistence upon which has ever been, and must continue to be, a necessary preliminary to human advancement.

3

CHAPTER III

THE RENAISSANCE IN SCIENCE

I. THE RENAISSANCE

IN the last chapter was described the manner in which the wisdom of the ancient philosophers had been passed on, through the channels of Arabic learning, to mediæval times.

Broadly speaking, the philosophers of the Middle Ages created nothing for themselves. Their science was not of their own production. It was almost entirely the science of the ancients, in the somewhat distorted and garbled forms in which it had been passed on to mediæval Europe through the vicissitudes of the intervening ages. Why should there have been this total lack of originality on the part of the philosophers of the Middle Ages ? Does it not seem a little remarkable that throughout Europe, and over a period of time of several hundreds of years, there should have been hardly anybody with anything new to offer to the world ? Yet there is an answer to this. We have but to remember the circumstances under which the reign of the ancient scientists gave way to the profound and universal ignorance of the Dark Ages—the decline and fall of the Roman Empire, the devastating penetrations of the Saracen soldiery, and the narrowing restrictions of Church discipline. A world of nations cannot emerge to the fullness of wisdom from intellectual darkness in a hurry. A keen appreciation of the wisdom of the ancients as it appeared from the efforts at reconstruction was all that was possible. The idea of criticizing what these ancients taught never so much as existed. And so there developed, quietly but effectively, a tradition of the infallibility of the old Greek philosophers which amounted almost to hero-worship.

We have next to consider how the world of mediæval science and philosophy came to an end, and how there was created in its stead that attitude of independence, inquiry, and experiment which may justly be regarded as characteristic of the modern period of science. The great movement that made for what we might call the evolution of modern science was the Renaissance. This was, in a sense, the sequel to that work of the reconstruction of the knowledge of the ancients to which we have referred, and it certainly found its inspiration in their example in literature, in art, and last of all in science.

Generally speaking, it may be said that although a compromise had been effected between the Church and classical teaching, mediæval learning in Science prior to the Renaissance was characterized by a subservience of thought to the theological purpose. The outlook as a whole was circumscribed. The astrological implications of the doctrine of macrocosm and microcosm were sufficiently complete to be satisfying, and so no attempt was made to go outside it. But now, at last, a spirit of unrest was beginning to manifest itself. The Portuguese navigators were showing not only a spirit of adventure which was in itself contagious, but were also providing new avenues of trade. As a result there was greater commercial prosperity. Thus people had more opportunity for thought; an opportunity, moreover, that was being aided by the era of printing that had just begun. Again in the intellectual world the physical speculations of Albert of Saxony and of Nicholas of Cusa, the geological investigations of Palissy, and the new and reliable versions of Ptolemy's *Almagest* that had been made available by Peurbach and Regiomontanus, were all beginning to have their effect. The factors in the scientific Renaissance are not, however, complete with-

out reference to the stimulus received in the direction
of improved technological processes.   These resulted not
alone from the commercial prosperity of the times, but
also from the stern necessities of an ever-present warfare.
All these considerations came into operation at a time
coincident with the conflict between the Protestants and
the Church of Rome, and brought into being that great
change in scientific outlook which led to the magnificent
achievements of such men as Copernicus, Kepler, Galileo,
and later Newton.

II. THE INFLUENCE OF NICHOLAS OF CUSA

These changing conditions were not rapid in their
advent.   The progress was slow and gradual, halting
even. Yet they were definite enough, and indeed began to
manifest themselves early in the fifteenth century through
the teachings of the illustrious German philosopher and
divine, Nicholas of Cusa, who, fittingly enough, was born
in the first year of the fifteenth century in 1401.   At an
early age, he was sent to the school of the Brothers of
Common Life at Deventer, in Holland.   In 1417 Nicholas
proceeded to Padua, and here he remained for six years,
primarily for the study of canon law.

Having graduated at Padua as a Doctor of Canon
Law, Nicholas entered the Church in 1425.   For the
next three years he studied divinity at the University
of Cologne, and shortly after he began a career of
ecclesiastical diplomacy and affairs which kept him
wandering over one part and another of Europe until
his death.   For many years he was engaged on various
missions as Papal Legate, and in 1460 his activities
brought him into conflict with Sigismund, Duke of
Austria, who, in defiance of the Pope (now Pius II),
imprisoned and ill-treated Cusa.   From this ill-treatment
Nicholas never recovered.   He escaped to Rome, and
afterwards resumed his travels on Church business.   He
died at Todi in Italy in 1464 in the presence of his old
friend Toscanelli the Geographer.

Such was his life. It is remarkable that amidst all the wanderings he should have found so much time for philosophical and scientific speculation. He was, however, essentially a thoughtful man, and he wrote extensively. The basis of his views was metaphysical. He was profoundly interested in all matters of observation and experiment, but this interest was at all times subservient to the larger metaphysical purpose. Standing in the forefront was his discussion on the movement of bodies, outlined in his *De docta Ignorantia*, written between 1439 and 1440, in the course of which he initiated an attack on the mediæval standpoint of a fixed earth which, willy-nilly, persisted throughout the next two centuries, and led ultimately to the enunciation by Copernicus of his famous system as an hypothesis, to the definite declaration of this system as a conviction by Bruno, and to its final establishment on mathematical grounds as a truism by Newton.

But there was another aspect of this inauguration— the institution of a definite experimental bias in philosophical inquiry. Not only do we find traces of this in his *De docta Ignorantia*, but we find it in full swing in the *De Staticis experimentis*, the fourth book of a series of papers entitled *Idiotæ libri quatuor*. In this Cusa gives his fundamental ideas on the use of the balance in medicine and in science generally. He quotes Vitruvius, recently rediscovered by Poggio, and gets ideas therefrom for such problems as that of the estimation of the speed of ships—a problem, incidentally, which later fascinated both Leon Battista Alberti and Leonardo da Vinci. Throughout, Cusa's contention is that through accurate comparisons by weight, various physical facts and properties are capable of investigation. So he suggests the comparison of waters from different springs, or water from the same spring at different times, of the blood and urine from old and young men, or of the same man in health and in sickness, and so on : suggestions which later led directly to Sanctorius' work on metabolic studies and to Van Helmont's gravimetric

analysis of urine. Another suggestion, virtually on
plant respiration, constitutes in effect the first biological
experiment of modern times, and offers the first formal
proof that the air has weight.

So we find in Nicholas of Cusa, in spite of the burden
of mediæval theology which he carried throughout his
career, the first fifteenth-century philosopher with a
truly modern outlook. We must under-estimate the
importance neither of his work nor of his influence.
He was the starting point of the Renaissance in science
in many a direction. In the world of philosophy he
was the forerunner of an illustrious line of thinkers,
from Pomponazzi and Ramus to Francis Bacon and
Descartes ; in his conceptions of the nature of matter
he foreshadowed the work of Paracelsus, and so led to
the dawn of modern chemistry ; and in astronomy he
was the first of a line which led through Peurbach and
Regiomontanus and Paul of Middleburg to Copernicus
and Kepler.

III. THE PHYSICS OF LEONARDO DA VINCI

Thus we see how Nicholas of Cusa pointed the way.
From this time onward progress in science was established. In the hands of Galileo in Italy and Gilbert in
England it was to receive an impetus in the creation
of experimental science that was to carry it forward in
triumph right up to the culminating period of Newton;
but meanwhile we have to note yet another figure of
scientific prophecy in the fifteenth century. From the
nature of his contributions to practically every branch
of scientific inquiry in his day, the name of Leonardo
da Vinci (1452–1519) stands out pre-eminently. Living
as he did at the crest of the wave of the Italian
Renaissance, he practically embodied in his being the
full expressions of its manifold activities. It is only
within comparatively recent years that the vast collection
of notes and sketches accumulated by Leonardo da
Vinci has been given the attention it deserves. Un-

fortunately, circumstances were such that after his death they were lost sight of, and it was only after the lapse of centuries that they once again came to light. This loss was a serious misfortune to science. Leonardo's work was in itself so fruitful and varied, and his outlook on nature was so essentially modern, that if only those who followed after him could have had access to his writings, and to his many anticipations of later discoveries in different fields of intellectual activity, there is no doubt that the course of scientific history would have been materially different in a number of important directions.

Leonardo da Vinci was born in 1452. At the age of fourteen he was apprenticed to the famous Florentine artist, Andrea Verrochio. From this time onwards Leonardo was constantly moving in scientific circles. Verrochio himself was a man of wide interests, which included a love of both geometry and perspective, and this naturally brought him into contact with many contemporaries of a like calibre. From all of these da Vinci must have learnt. Consequently when Leonardo left Florence to come to Milan in 1483, the seeds of a scientific career had been truly well sown. Leonardo's position at Milan was that of a consulting engineer to Ludovico Sforza. In this capacity he was brought continuously into contact not only with the scientific and technical problems of architecture and of military and civil engineering, but also with a number of illustrious scientific contemporaries of the period.

Towards the year 1499 political events began to make Leonardo's position at Milan untenable, and in December of that year he set out, accompanied by his friend Pacioli, for Florence. The next few years found him alternating between scientific and engineering problems on the one hand, and incompleted works of art on the other, until in 1507 he returned to Milan as painter and engineer-in-chief to the French King Louis XII. In the summer of 1515 he accepted an invitation from Francis I of France, Louis XII's successor, to take up his residence

in the Castle of St. Cloux, near Amboise. Here he spent
the remainder of his days. He died on 2nd May, 1519.

Leonardo very happily combined within himself those
excellent qualities which produce both the theorist in
science and the technologist who blends his theory with
practice, and what is equally important, his practice
with theory.

Formal mechanics occupied a large share of his time
and thought. In it, as in science generally, we find in
Leonardo every evidence of that spirit of independence
and experimental inquiry which gave to Galileo, a
hundred years after him, the title of ' Father of Experimental Science.' Galileo deserved this title, but it was
accorded him in ignorance of the labours of da Vinci.
It neither detracts from his glory nor does injustice to
his forerunner, therefore, if we plead for Leonardo the
corresponding title of ' Grandfather of Experimental
Science.'

He lovingly referred to the study of mechanics as
'The paradise of the mathematical sciences.' In this
subject he opened up many new fields of investigation
hitherto untouched. He knew of the principle of inertia.
He tells us that 'no body can move of itself, but by the
action of some other, and that other is force'; of moving
bodies, too, he tells us that ' all movements tend to
maintenance'; whilst what he knew of the law of
reactions is clear from his reference that 'an object offers
as much resistance to the air as the air does to the
object.' His study of falling bodies is interesting from
several points of view. We may summarize his views
as to falling bodies by the following quotation: 'Why
does not the weight remain in its place ? . . . Because
it has no resistance. Where will it move to ? . . . It
will move towards the centre of the earth. And why by no
other line ? . . . Because a weight which has no support
falls by the shortest road to the lowest point, which is
the centre of the world. And why does the weight know
how to find it by so short a line ? . . . Because it does
not depend and does not move about in various direc-

tions.' Here we get Leonardo's philosophy of the falling body in a nutshell, so to speak.

Not the least interesting of Leonardo's mechanical researches concern themselves with the principle of work. He did not, of course, use the term work. He did, however, appreciate the fact of a value in, and a measure of what we may speak of as the ' achievement ' of a force. Thus he writes that ' if a force carries a weight in a certain time through a definite distance, the same force will carry half the body in the same time through double the path.' He recognized, in effect, a definite limit to the results of a given effort ; and also that this effort was not alone a question of the magnitude of the force, but of the distance through which it acts. If the one be increased, it can only be at the expense of the other.

Intimately linked up with this principle of work was the age-old myth of perpetual motion. If the principle of work be true, then the achievement of motion is impossible. On this matter Leonardo had no illusions whatever. There were, however, many contemporary with him who thought otherwise, and with these da Vinci had no patience. ' Oh speculators on perpetual motion,' he writes, ' how many vain projects of like character you have created. Go and be the companions of the searchers after gold.' Leonardo's dynamics also included studies of motion down an inclined plane, and the collision of bodies. When, again, we turn to statics, we find an even wider range of scientific activity on the part of our philosopher. We must remember, however, that here he was treading on less virgin soil. The works of Aristotle, Archimedes, Euclid, Hero, Pappus, and others during the Greek era, and of Jordanus Nemorarius, Albert of Saxony and others in the Middle Ages, were known to him. On their foundation, however, he built very securely and his notes show clearly and conclusively that he fully understood the lever principle, centres of gravity, pulley systems and their mechanical advantage, and many other important

branches of modern statics. We may sum up, in fact,
by saying that the theoretical basis of his work in
mechanics was not only sound but was in extent far
beyond his times.

Turning next to the group of subjects of heat, sound,
light, magnetism, etc., collectively referred to nowadays
as physics, we come to a wide sphere of activity on the
part of Leonardo da Vinci. His notes on these subjects
are scattered very generally through his various note-
books, but as might naturally be expected, the subject
which comes in for most of his attention is that of
optics. On the other hand, although the earliest references in history to the simple phenomena of electrification
by friction carry us back to Thales of Miletus, there
appears to be no mention whatever in Leonardo's
manuscripts of this subject.

The references to heat are somewhat scanty, but in
any case very little had been done by philosophers,
apart perhaps from Roger Bacon in the thirteenth century, towards serious discussion on this subject. Leonardo appears to have linked up the phenomena of
heat with light from the point of view that heat springs
from luminosity, and he instances such varied examples
as the sun and the flame of a candle. The expansion
property is evidently referred to as the ' greater solidity
of the fluids where there is a greater coldness.' His
observation is at fault, however, in a sketch which
shows a thick metal slab heated in front of a fire and
bent inwards through expansion. The back surface
being the cooler the expansion must, of course, be
outwards.

Although Leonardo may not have understood the
theory of convection currents, he certainly knew of
their application, and a number of interesting sketches
are to be found of various roasting spits—at this time
a very popular device. He observes by way of introduction to this that when two equally heavy objects
are suspended from two sides of a balance, and one
of them be heated, the hot air in rising will carry it

upwards and cause the colder body to descend. Leonardo shows how air treated by a fire below rises and causes a suitably vaned wheel at the top of the chimney to rotate. The meat is suspended from mechanism geared to the vane, and so rotates also.

Leonardo's notes on magnetism are not very plentiful, but are none the less interesting. At this time both the magnetic properties of the lodestone and the directive property of the compass were well known. Leonardo shows in his sketches that he agreed with the current notion that the compass property of a magnet was caused by the attraction of the Pole Star. Thus he writes : ' Take a large vessel and fill it with water. In this water place a wooden vessel and in it place a magnet without any other water. The magnet will float like a ship and immediately after its power of attraction will cause it to move towards the Pole Star.' We may with some confidence ascribe to Leonardo the invention of some scheme of ring suspension for the purpose of maintaining the ship's compass in a horizontal position in spite of the movements of the ship. A sketch also seems to indicate that Leonardo knew of the phenomenon of magnetic dip. One sees a vertical board spiked at the base so that it may be pushed into the ground. At the upper part of the board is a graduated circle carrying a horizontally mounted magnetic needle sketched to suggest an angle of dip. A plumb line is attached to the apparatus.

We come next to Leonardo's notes on sound. They are remarkable chiefly for the suggestions of some sort of relationship between the mechanisms of propagation of sound and light, and also for the references to some conception of wave motion. Of the origin of sound he writes that ' A note could be called forth in a body by means of a blow, and this blow could only be made through a movement.' From this he passes to the problem of echoes, and he asks the question ' If the echo of a voice answers me twice at a distance of thirty yards with two degrees of the power of the noise, with how many degrees of the power of the noise will it answer me at a distance of one hundred yards ? '

There is no doubt that Leonardo appreciated the fact that sound spreads out in spherical form accompanied by a diminution in intensity. However vague his notes undoubtedly were, it is clear he had some idea of a wave motion that was comparable with the spreading out of ripples in circles when a liquid surface is disturbed. Thus we read :

' Because in all cases the movement of water has great similarity with that of air, I will connect the example with the above named principles. I say, if you throw two little stones into a lake of still water at the same time and at a little distance from one another, you will see called forth round about the two situated points of contact two separate groups of circles, which groups grow and finally meet each other . . . and the principle of this is that although there is shown an appearance of movement the water itself is not moved from its place, because the openings which are made by the stones at once close themselves again and cause a certain disturbance to be set up, which one can speak of rather as a trembling than as a movement.'

CHAPTER IV

THE DAWN OF EXPERIMENTAL PHYSICS

I. THE SIXTEENTH CENTURY—WILLIAM GILBERT

WITH the advent of the sixteenth century came the culmination of the Renaissance in science —the application of the test of experiment. We have seen how the dawn of the fifteenth century produced the herald of this revolution in scientific method in Nicholas of Cusa, and how it found a brilliant exponent in Leonardo da Vinci. Leonardo, however, held no academic appointment, and therefore gave little public utterance to his precepts. His wonderful mass of notes, too, was practically lost at his death, and has only come to light within the last hundred and fifty years. Nevertheless, the general impetus to the new spirit of inquiry and to the growing tendency to question the infallibility of ancient authority steadily grew, and two exponents of the new methods stand out in the sixteenth century as the founders of experimental physics in its real modern sense. These are William Gilbert in England, and Galileo Galilei in Italy.

William Gilbert was the son of an old Suffolk family who had migrated to Colchester, and here he was born in 1540. He went to St. John's College, Cambridge, in 1558. For some years he studied and lectured in mathematics, but later took up medicine, graduating as M.D. at the age of twenty-nine years. After some years of travel he established himself in a medical practice in London, and for some time held the appointment of physician-in-ordinary to Queen Elizabeth.

His house on St. Peter's Hill, near St. Paul's Cathedral, became the rendezvous of a number of men of science who formed themselves into a society or college of philosophers—an unofficial organization that was to prove the forerunner of such later institutions as the Royal Society. Its formation was at least evidence of the high esteem in which Gilbert was held as a man of science. Gilbert died in 1603.

II. MAGNETISM AND ELECTRICITY PRIOR TO GILBERT

The book that embodied Gilbert's researches and that constitutes the memorial to his greatness was entitled : *De Magnete, Magneticisque corporibus et de magno magnete tellure* (' On the magnet and magnetic bodies, and on the great magnet, the earth '). It ranks as a document of first importance in the history of science, and it will be advisable, before discussing it, to review in brief such progress as had been made in the subject of magnetism and electricity prior to Gilbert's time.

We have seen that the property of attraction possessed both by magnetite for iron and by amber when rubbed for very light substances was known to the ancients. The directional property of a suspended magnet, however, was not known before the eleventh century. It is probable, however, that the Chinese also knew of the phenomenon of magnetic declination (i.e., of the fact that a suspended magnet does not point due north and south, but deviates from this direction by a few degrees). One Chinese writer tells us that ' the soothsayers rub a needle with the magnet stone, so that it may mark the south ; however, it declines constantly a little to the east. It does not indicate the south exactly.' This compass property of the magnet had not only become common knowledge by the thirteenth century, but it was also attributed to an attraction somehow set up by the Pole Star—a view to which we have seen even Leonardo da Vinci subscribed.

An early contributor to progress in the knowledge of the compass was Peter Peregrinus de Marincourt, a man who greatly influenced Roger Bacon in the direction of scientific research. Taking a piece of lodestone that had been rounded off into a globular form, he placed a needle on it at a large number of points round the globe, and charted the directions in which it set. He found that the directions lay along a number of 'great circles' that all passed through two opposite points. These he referred to as the ' poles ' of the magnet. He definitely refers to the polarity property of all magnets, and gives the law that like poles attract each other. He also states that if a magnet be divided into two, each fragment has two poles. Peregrinus also invented a compass with a graduated scale and a pivoted needle, and by the sixteenth century all ships carried compasses mounted on compass cards.

III. GILBERT'S *DE MAGNETE*

There are two aspects of the work of William Gilbert that mark him out as a man of eminent service to civilization. The first is his inauguration of the experimental method and a repudiation of the blind subservience to ancient authority, and the second is his establishment of specific facts regarding terrestrial magnetism. With regard to the former of these two contributions we will let him speak for himself. In his preface he writes :

' Why should I submit this . . . new philosophy to the judgment of men who have taken oath to follow the opinions of others, to the most senseless corrupters of the arts, to lettered clowns, grammatists, sophists, spouters, and the wrong-headed rabble, to be denounced, torn to tatters, and heaped with contumely. To you alone, true philosophers, ingenious minds, *who not only in books but in things themselves look for knowledge*, have I dedicated these foundations of magnetic science—a new style of philosophizing.'

An example of his appeal to experiment as a new departure in scientific method is well afforded by the following passage :

'In the discovery of secrets and in the investigation of the hidden causes of things, clear proofs are afforded by trustworthy experiments rather than by probable guesses and opinions of ordinary professors and philosophers. In order, therefore, that the noble substance of that great magnet, the earth, hitherto quite unknown, and the exalted powers of this globe of ours may be better understood I shall first of all deal with common magnets, stones, and iron materials, and with magnetic bodies, and with the near parts of the earth, which we can reach with our hands and perceive with our senses. After that I shall proceed to show my new magnetic experiments, and so I shall penetrate for the first time into the innermost parts of the earth. . . .'

Let us now turn to the experiments that earned for him the name of ' Father of the Magnetic Philosophy.'

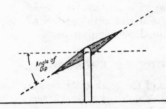

Enlarging first upon the observations of Peter the Pilgrim referred to above, Gilbert refers to the lode-stone made into a globe ' upon the turning tool used for rounding crystals ' as a *terrella* (little earth) and describes

FIG. 2.—The Angle of Dip

how with the aid of either a wire or a *versorium* (a piece of iron shaped with an arrow head and mounted on a pivot, i.e., the equivalent of the modern compass needle), the ' meridian circles ' could be charted out and the magnetic poles ascertained. He then shows how to distinguish between the *austral* (N-seeking) and the *boreal* (S-seeking) poles by mounting the terrella on a float upon still water, and noting the direction in which it settles. Then, by bringing up to this another magnet whose poles have been previously determined, he proceeds to demonstrate how, if the ' austral ' pole is held ' towards the boreal pole of the one that is swimming, the floating stone forthwith follows the other stone, and does not leave nor forsake it until it adheres,' whereas ' if you apply the austral to the austral, the one stone puts the other to flight.' In other words, he clearly establishes that *unlike poles attract each other, and like poles repel each other.*

Gilbert now sets himself to explain the compass-property of magnets—i.e., why they always point in one direction when freely suspended. His intuition leads him correctly. Certain resemblances between the earth and the proved facts of his terrella become charged with significance. ' The lodestone has poles, not mathematical points, but natural termini of forces excelling in primary efficiency by the co-operation of the whole. Now there are poles in like manner in the earth which our forefathers sought ever in the sky; for the sky has an equator, a natural dividing line between the poles, just as the earth has.' So he advanced to the view that the earth is itself a giant magnet, and thus was at once able to explain, by the operation of the law that like poles repel and unlike poles attract, that a suspended needle sets it-self so that its N-seeking

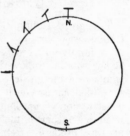

Fig. 3.—Gilbert's Terrella

pole is pointing to the earth's opposite magnetic pole in the region of the geographical North Pole. To this fact of the earth's magnetism he was also able to at-tribute the explanation of the phenomenon of *magnetic dip* which had by then become known. When a needle is magnetized and is then suspended at its centre on a horizontal axis in the plane of the magnetic meridian (i.e., the plane passing through the earth's magnetic pole), it swings round and takes up an inclined position (Fig. 2) and the angle with the horizontal is known as the *Dip Angle.* Gilbert knew that this angle varied in amount from zero at the poles to 90° at the ' mag-netic equator '; and he got similar results with his versorium in relation to the terrella. The explanation

4

is that from any given point on the earth (or terrella),
the versorium is acted upon by a force of attraction,
say, towards the earth's magnetic N-pole, and a force
of repulsion from the earth's S-pole, and the needle
therefore takes up a resultant intermediate direction
whose value must vary from place to place (Fig. 3).

Gilbert's *De Magnete*, published in London in the
year 1600, was thus a work of first magnitude, and it
is interesting to note in conclusion that it was the first
purely scientific work of importance to be published in
this country.

IV. GALILEO GALILEI

The great Italian contemporary of Gilbert, Galileo
Galilei, was born at Pisa in 1564, and was the eldest
of a family of three sons and three daughters. His
father was an impoverished nobleman of wide culture
who wrote on music.

Galileo as a youth soon exhibited a bent for the
construction of mechanical models. He also inherited
from his father a taste for music and drawing, and
it was probably from a desire to understand the
mysteries of the theory of music and of perspective that
he was drawn to a study of mathematics and physics.
In 1581 he entered the University of Pisa as a medical student. It was during his student days that he
designed the famous ' pulsilogium '—virtually a simple
pendulum of adjustable length for the timing of pulse-
rates in patients. The story is well known, and is
important for the discovery of the *principle of isochronism* that was involved, and that forms the starting point
of modern horology. He was in the cathedral during
a service and noticed a swinging chandelier. Galileo's
attention was caught by the fact that although the
amplitude or sweep of the swing was dying down, the
time of the swing appeared to be constant. Hence his
invention of the pulsilogium.

Galileo's taste for Geometry led him instinctively

to the works of the ancients. From Euclid he passed
to Archimedes, and it was this latter philosopher's
work on floating bodies that inspired him to an investigation on the hydrostatic balance, and later on the
subject of centres of gravity. His essays on these
subjects brought him to the notice of the Duke of
Tuscany, and as a consequence he was appointed in
1589, in his twenty-sixth year, to the professorship of
mathematics at the University of Pisa.

It was not long before, in the ardour of
his youth, he initiated his first attack on
ancient authority. In particular he singled
out Aristotle's famous doctrine that a heavy
body falls freely from rest faster than a light
body, and in his classic experiment from the
top of the famous leaning tower at Pisa, he
proved in so many seconds to an astonished
and an all-unwilling audience that, allowing
for the resistance of the air, the Peripatetic
was wrong. In the simultaneous crash of
a heavy and a light weight, those who, in
the words of our philosopher, 'fancied that
science was to be studied like the Æneid
or the Odyssey, and that the true reading
of nature was to be discovered by the colla-
tion of texts' were both confounded and
dumbfounded. It was an unpopular victory
and the beginning of a tragedy for Galileo.

Fig. 4.—
The Prin-
ciple of
Galileo's
Thermo-
meter

In an immediate sense it led to an un-
popularity among his colleagues that prompted him
to leave Pisa in 1592 for a professorship at Padua. It
was at Padua that Galileo invented his Thermometer
—a glass tube terminating in a bulb, and inverted with
the open end dipping into a small flask containing
liquid. Some of the air within the bulb having been
withdrawn, a column of the liquid rose in the tube,
and the height of this column (Fig. 4) varied with the
temperature. The instrument was fairly sensitive to
small temperature changes, but though Galileo did not

know it, the height of the column was also affected by changes of barometric pressure.

During the course of Galileo's stay at Padua, he continued his experiments on falling bodies, but we will defer a consideration of the details until later in the chapter. It was at this time, too, that he became fully convinced of the truth of the teachings of Copernicus, and accepted the heliocentric theory of the universe in opposition to the geocentric doctrines inherited for centuries from Ptolemy. However, in spite of this inward conviction, Galileo continued to teach the Ptolemaic system to his students. It was a politic attitude to adopt. The prejudices of the times and the discipline of the Church had come sharply into evidence about that time when Bruno had been burnt at the stake for his open pronouncements in favour of Copernicus. So Galileo chose the course of prudence, but only up to a point. The Aristotelian doctrine of the immutability of the skies had long irked him, and the advent of a new star in 1604 gave him an opportunity that he could not resist, and the popularity of his criticisms among the students was such as to throng his lecture-rooms with auditors. It was a significant trick of fate, moreover, that brought to him rumours in 1609 of the invention of the telescope from Holland, and without any details before him, Galileo thought the problem out for himself and produced an instrument which proved the final weapon for the unmasking of the fallacies of the ancients. Boldly proclaiming for the Copernican doctrine, he proved conclusively by the aid of a succession of astronomical discoveries made possible with his telescope—Jupiter's moons, Saturn's rings, spots on the sun, mountains on the moon, and others—that the immutability of the skies was a chimera.

In 1610 Galileo accepted the post of mathematician and philosopher to the Grand Duke of Tuscany, and left Padua for Florence. It proved to be a fateful move. Padua was in the Republic of Venice, and

enjoyed some measure of immunity from the ecclesiastical authority of Rome. Here scepticism of Church
doctrine was tolerated. But Florence was differently
circumstanced, and he who denied the earth as the
centre of the universe courted trouble. In 1615 Galileo
was summoned to Rome, and after a long inquiry he
was ordered, on 26th February, 1616, under threat
of imprisonment and torture, ' to abandon, and cease
to teach his false, impious, and heretical opinions.'
Galileo was powerless. He gave his promise and
returned to Florence. The succeeding years proved
but a respite. He continued his researches, and later,
falsely encouraged to a feeling of security by the elevation to the Papacy of one who had been his friend, he
returned to the inevitable Copernican theme and wrote
his *Dialogues on the Ptolemaic and Copernican Systems.*
The book appeared in 1632. Its effect was immediate.
The anger of the Church burst in full flood. Galileo,
now an old man, was once more summoned to Rome.
He arrived in February of 1633, and the proceedings
of the Inquisition, conducted in secret, were concluded
by June of the same year. It was a long enough interval
of mental torture, and it saw its culmination in the
historic scene of recantation when our aged philosopher,
garbed in penitential clothing, was forced to make his
famous utterances before the assembly of cardinals.
' With a sincere heart and unfeigned faith, I abjure,
curse and detest the said heresies and errors. . . .
And I swear that I will nevermore in future say or
assert anything verbally, or in writing, which may give
rise to a similar suspicion of me.'

Thus, broken and grief-stricken (the loss of a beloved
daughter adding to his sorrows), Galileo was permitted
to end his days in some sort of open confinement at
Arcetri. That he was by no means destroyed as a
scientific force is, however, clear from his famous
Dialogues on Motion—a brilliant study on dynamics
based on experiments at Arcetri, and published in
Amsterdam in 1636. In his last years he was stricken

with blindness, and he died in 1642 at the age of seventy-eight years.

V. GALILEO'S PHYSICS—MECHANICS

Galileo's contribution to physics, with which alone we are here concerned, falls into two categories—his researches in mechanics and his construction of the telescope. We will here consider his mechanics. We may say of him that he virtually founded the study of dynamics. We have seen that in fact Leonardo da Vinci carried out many important investigations in this branch of physics, but unfortunately his work was lost to the world. The ground he covered had therefore to be re-traversed in ignorance of his work. This was a service adequately performed by Galileo.

Let us see as a preliminary what was the position of dynamics as he found it initially. The prevailing ideas were of course a heritage of Aristotelianism. The falling of heavy bodies and the rising of light bodies (e.g., liquids) was explained by the assumption that every object sought its *place*. The place of heavy objects was below, and that of light objects was above. Motions were classified as *natural* (e.g., falling) or *violent* (e.g., projectiles). It was also assumed that heavy bodies fall more quickly than light ones.

Galileo's earlier researches on falling bodies were recorded in his *Discorsi e dimostrazione mathematiche* in 1638. They show the scientific spirit from the outset. There is no attempt to consider *why* bodies fall. The sole concern is *how* and by what law ? Galileo's line of approach is full of interest to students of scientific method. Observation showed him that a falling body gains in velocity as it falls. He therefore started with an assumption that the increase in velocity (i.e., the acceleration) is proportional to the distance moved, and proceeded to put this assumption to the test of argument and experiment. As we now know, the assumption certainly was incorrect, but curiously enough Galileo satisfied himself as to its falsity by a train of reasoning that was in itself fallacious.

'If the velocity with which a body overcomes four yards is double the velocity with which it passed over the first two yards, then the times necessary for these processes must be equal; but four yards can be overcome in the same time as two yards only if there is an instantaneous motion. We see, however, that the body takes time in falling and requires, indeed, less time for a fall of two than of four yards. Hence it is not true that the velocity increases proportionally to the distance fallen.

Having rejected this initial assumption, Galileo now makes a second one. This time he assumes the velocity acquired to be proportional to the time taken. His inquiry on this basis is as follows : Let the line O A (Fig. 5) be a time axis along which successive portions represent time intervals of descent. At the ends of these portions vertical lines are erected proportional to the velocity acquired. Thus H G represents the

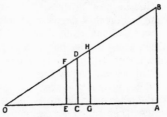

FIG. 5.—To illustrate Galileo's Investigations in Dynamics

velocity acquired corresponding to the time of descent O G. Consider the point C at half the total time of descent. The velocity C D will be one-half the final velocity A B if the initial assumption is correct. If now we consider two further points E and G at small equal time intervals before and after C, it will be seen that the velocity H G is as much greater than the mean velocity C D as E F is less; and that remains true for every such pair of points before and after C. The motion on this law of increase has therefore the same result as would a *uniform* motion with the speed at C (i.e., half the final velocity) since the total loss suffered by the slower speeds in the first half is exactly counterbalanced by the total gain in the second half.

Hence if v denote the terminal velocity, $\dfrac{v}{2}$ is the equivalent uniform velocity to cover the same ground s in the same time t, whence $s = \dfrac{vt}{2}$. But if g denotes the speed acquired after falling for unit time (i.e., the acceleration), then in time t the speed, from the initial assumption, will have become $v = gt$. Hence the distance $s = \dfrac{vt}{2}$ becomes $\dfrac{(gt)t}{2}$, i.e., $s = \tfrac{1}{2}gt^2$. In other words, Galileo was led to the deduction that *the uniformly accelerated motion must be such as to make the distance traversed proportional to the square of the time.*

This he now proceeded to put to the test of experiment. For this purpose he used as an inclined plane a board about twelve yards long down the centre of which was a groove one inch in width, lined with smooth parchment to minimize friction. At distances from the upper extremity proportional to the numbers 1, 4, 9, 16, etc., marks were made in anticipation of their equivalence to time units, of descent of 1, 2, 3, 4, etc. Galileo then timed the descent of a highly polished brass ball down the groove, and did indeed find his assumption fully borne out by the results, namely that the distance of descent was proportional to the square of the time. He repeated the experiment over a varying range of some hundred different inclinations of the plane. A serious difficulty that confronted him was the accurate measurement of time. This he very ingeniously overcame by arranging a large vessel of water with a small hole at the bottom, closed with the finger. This was released at the exact instant of the release of the ball, and the hole was closed again at the exact termination of the experiment. The escaping water was caught in a balance and weighed and the weight was naturally a measure of the time of descent of the ball down the plane.

Finally, it remained for Galileo to link up the results

for motion down an inclined plane with the case of a freely falling body.   His argument was as follows :

' If a body falls freely downwards, its velocity increases proportionately to the time.   When then the body has arrived at a point below, let us imagine its velocity reversed and directed upwards ;   the body then, it is clear, will rise.   We make the observation that its motion in this case is a reflection, so to speak, of its motion in the first case.   As then its velocity increased proportionately to the time of descent, it will now, conversely, diminish in that proportion.   When the body has continued to rise for as long a time as it descended, and has reached the height from which it originally fell, its velocity will be reduced to zero. We perceive therefore that a body will rise, in virtue of the velocity acquired in its descent, just as *high* as it has fallen.

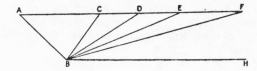

Fɪɢ. 6.—To illustrate Galileo's Investigations in Dynamics

If, accordingly, a body falling down an inclined plane could acquire a velocity which enabled it, when placed on a differently inclined plane, to rise higher than the point from which it had fallen, we should be able to effect the elevation of bodies by gravity alone.   There is contained, accordingly, in this assumption, that the velocity acquired by a body in descent depends solely on the *vertical* height fallen through and is independent of the inclination of the path, nothing more than the uncontradictory apprehension and recognition of the *fact* that heavy bodies do not possess the tendency to rise, but only the tendency to fall.'

Galileo did as a matter of fact proceed to experiments to amplify this perfectly logical train of reasoning, but we do not need to follow him further.   It merely suffices to point out, with Mach, that he did more than supply the world with a *theory* of falling bodies—he definitely established the *facts* of falling.   We will point out one further observation of Galileo's on account of its profound importance as subsequently developed by Huyghens and Newton.   He considered a body falling down the inclined plane A B (Fig. 6). With its final velocity at B it now ascends one of any such planes as B C, B D, B E, etc. Clearly since at any of the points C, D, E, etc., its final velocity will be zero, and as the paths B C, B D, B E, etc., become successively longer, the *retardations* along B C, B D, B E, etc., are successively smaller as the planes more and more approach the horizontal, and in the limit along the horizontal plane B A the retardation must be zero (assuming there is no friction of course). In such a case the body will move indefinitely with constant velocity. Galileo thus developed the so-called law of inertia which formed the basis later of Newton's famous laws of motion, namely, that a body not under the influence of forces (i.e., of special circumstances that change motion) will continue to move for ever with the same velocity in the same direction.

CHAPTER V

THE PHILOSOPHERS OF THE ATMOSPHERE

I. THE DOCTRINE OF THE *HORROR VACUI*

THE story of physics has now reached its essentially modern stage. With the advent of the two giant personalities in science of the sixteenth century whose work we have just described, the experimental method had come to stay, and from now onwards our story becomes one of continuous advance, step by step, haltingly at times perhaps, and feverish in its rapidity at other times, but at least continuous at all times. The new method was all that was required to open up vast possibilities for the future, and the results were quick to show themselves. The seventeenth century, indeed, is in some respects the most brilliant in the history of physics. It brought into being a wonderful galaxy of philosophers, including the renowned Sir Isaac Newton, and some of its members it is now our purpose to consider.

One of the earliest results of the new methods of inquiry was to bring to light the true facts as to the physics of the atmosphere, and since the atmosphere plays so vital a part in our lives, it is fitting that we devote a special chapter to the very important series of researches connected with it.

Up to the time of the sixteenth century this subject, like so many others, was dominated by the old dictum of the ' horror vacui ' inherited from the ancients— Nature abhors a vacuum. It was a phrase characteristic of Greek philosophy and consistent with the general medi-

53

aeval outlook.   It possibly arose from the inherent
difficulty always encountered in obtaining a piece of
space with nothing in it.   The ancients, for example,
came into contact with the problem in connection with
the action of the pump—a mechanism with which they
were familiar.   When the piston is raised, it certainly
tends to produce a vacuum, but because of Nature's
abhorrence, water rose in the well-pipe to prevent the
vacuum from forming.   This was the view that received
the stamp of authority from Aristotle, and it was moreover a sufficiently plausible view to satisfy, and so the
idea persisted.   For the first step in the weaning of the
world from this view-point we have to return to Galileo.
The story goes that the reigning Grand Duke of Tuscany
had ordered a well to be dug to supply the ducal palace
with water.   The workmen employed thereon came upon
the water at a depth of 40 feet, and the next step was
accordingly to pump it up.   A pump was erected over
the well, and a pipe was let down to the water, but the
water was found to rise to a height of 33 feet and no
more, in spite of all efforts at the pump-handle and in
spite of the most careful overhauling of the pump
mechanism.   It was at this stage that Galileo was consulted.   While the famous philosopher was unable to
offer a solution, he at least indicated the problem.
Here above the 33 feet of water was 7 feet of ' vacuum '.
Clearly, then, if it were true that Nature abhors a
vacuum, there was a limit to its abhorrence, and this
limit appeared to be 33 feet.   Why should there be
this limit ?

II. TORRICELLI (1608–47)

The first step—and it was an important step—towards the solution was taken by the young friend of
Galileo's old age, Evangelista Torricelli.   He was born
at Faenza in 1608, and, left fatherless at an early age,
he was taken care of by an uncle, who sent him to Rome
in 1627 to study science at the Collegio di Sapienza.   He

was inspired by the reading of Galileo's works to a special study of mechanics, and wrote a work on the subject, *De Motu*, which so pleased Galileo that in 1641 the aged philosopher invited Torricelli to come and reside with him, and here he stayed until the great master died. Torricelli was now offered, and he accepted the post of grand ducal mathematician and professor of mathematics in the Florentine Academy, a post he held until his death from pleurisy in 1647 at the early age of 39 years.

His contributions to physics were mainly in the direction of mechanics, but they were important ones. They related to the mechanics of moving bodies, to general theorems on centres of gravity in relation to the equilibrium of bodies, and to certain theorems in hydrodynamics. His most famous contribution was, however, the invention of the barometer, and it is with this aspect of his work that we are here chiefly concerned. Torricelli was presumably introduced to the problem of the vacuum by Galileo. He hit on the idea of substituting mercury for water. His expectation was that as the density of mercury is nearly fourteen times that of water, the pump that would draw a column of 33 feet of water would only draw one-fourteenth such a column of mercury. Galileo had already proved that air has weight by first weighing a bottle containing air, and then again after much of the air had been expelled by heating. There was a measurable loss in weight. Therefore, argued Torricelli, the atmosphere has weight, and must therefore exert pressure ; and in his view the column of water drawn up by the pump ascends to such a height that its weight exactly balances the pressure of the atmosphere. In other words, the atmosphere, having weight, presses on the surface of the water in the well ; and when this pressure is removed from the water inside the pump-pipe by the withdrawal of the air on the action of the pump-handle, the liquid is forced up by the atmospheric pressure outside, until the weight of the water column within the pipe produces an equal counter-pressure. So Torricelli devised his famous experiment, and it was carried out in 1643 at his direction by his pupil Vincenzo Viviani (1622–1703). A strong glass tube (Fig. 7(a)) about 3 feet long was closed at one end and filled with mercury to the brim. The upper end was now closed with the thumb and inverted over a dish of mercury, and the thumb was then released. Although some of the mercury now ran out, there was left a column nearly 30 inches high (Fig. 7(b)). The column corresponded exactly to the water column of 33 feet, i.e., it was nearly one-fourteenth of the height of the water column. This clearly confirmed Torricelli's view that the column of mercury was sustained by the pressure of the atmosphere although he found difficulty in persuading some of his contemporaries that the space above the mercury in the tube was a vacuum. Finally, Torricelli noticed that his mercury column exhibited slight changes in height from day to day, and he did not hesitate to attribute these to changes in the pressure of the atmosphere.

Fig. 7.— Torricelli's Experiment

III. BLAISE PASCAL (1623–62)

Torricelli published no account of his experiments, probably because he was too busy. He did, however, describe them in letters to Ricci, a friend in Rome, in the course of which he said that 'the aim of my experiments was not simply to produce a vacuum, but to make an instrument which shows the mutations of the air, now heavier and dense, and now lighter and thin.' Ricci in 1644 communicated these results to Father Mersenne in Paris, and through him the news of the experiments reached a brilliant young scientist named Blaise Pascal. There were obviously no details available to Pascal of the experiments, since the lines along which he studied the problem show that he traversed the whole ground anew.

Blaise Pascal was born in 1623 at Clermont in Auvergne, and he exhibited signs of a remarkable precocity of genius at a very early age. In fact by the age of sixteen years he had completed a work on conic sections of such importance as to form the basis for the modern treatment of the subject. It included a number of theorems, the most famous of which has ever since been known as Pascal's Theorem. This states that the intersections of the three pairs of opposite sides of a hexagon inscribed in a conic section are collinear. Pascal himself spoke of this as the mystical hexagon.

In 1641, after a period of adversity, Blaise's father, Etienne Pascal, removed with his family to Rouen, and here Blaise engrossed himself in his studies so avidly as to impair his health very seriously, and he suffered from a form of paralysis. In spite of his weak health, however, he continued to take life gaily until the year 1654, when occurred that famous accident which produced his conversion to an ascetic and religious life. He was driving to Neuilly when the horses got out of hand and ran away, and there is little doubt that Pascal would have been killed but for the fortunate breaking of the traces. There are but scanty details of the later years of his life. His last long illness began in 1658, and he got steadily worse and died in 1662 in his thirty-ninth year.

A man of letters, a theologian, a philosopher, a mathematician, and a scientist, his was a brilliant record of achievement. In physics his experiments and his treatise on the equilibrium of fluids, written before 1651 but not published till the year after his death, entitle him to be regarded as one of the founders of the science of hydrodynamics.

Let us now see in what way Pascal contributed to the physics of the atmosphere. His first thoughts were directed to the vexed problem of the 'Torricellian Vacuum' above the liquid column in the tube. He accordingly repeated the experiment, using both mercury and red wine (which, being lighter than water, gave a column about 40 feet in length), and by inclining the tube he soon convinced himself that the space above was indeed a vacuum. This was an unpopular view in France, but Pascal indicated that a vacuum could

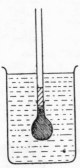

be produced by other means, as for example by a glass syringe the nozzle of which, being first closed by the finger under water, is then drawn back. No air could get in, and so a vacuum must be produced. These and similar experiments led Pascal to realize the complete analogy between the phenomena of liquid pressure and water pressure. Thus he showed that flies can withstand injury against high pressure in fluids so long as the pressure is equal on all sides, and he deduced from this similar facts regarding fish in the water and the animals living in the air.

FIG. 8.—Pascal's Experiment on Liquid Pressure in relation to Depth

Pascal's next observation led to important consequences. He took an open tube and tied a bag of mercury to its base. The whole was then slowly lowered into a deep vessel of water (Fig. 8), keeping, however, the top of the tube open to the air. Pascal found that the deeper the tube was sunk, the higher did the mercury rise *owing to the constantly increasing water pressure with the depth*. Pascal argued that the same reasoning must apply to the layers of atmosphere—namely, that the atmospheric pressure must be less at high altitudes than at low ones. Pascal was at Rouen at this time, and he decided to compare the barometer reading at the top of a church tower in the city with that at the base. The difference was there, but it was too slight, and Pascal realized he must seek wider comparisons. There being no high hills round Rouen, and his own health being too precarious to risk his travelling, he got into touch with a brother-in-law named Perier whom he knew to have scientific interests, and who lived in the mountainous district of Auvergne.

The mountain selected was the Puy de Dôme, near Clermont, where Perier awaited favourable weather opportunities. Here, on the morning of the 20th September, 1648, he started with some friends, and equipped with two glass tubes each 4 feet long. At the base the Torricellian experiment was performed, and the resulting mercury columns were both found to read 23 Paris inches and $3\frac{1}{2}$ lines. One was then fixed and left in the case and the other taken up the Puy de Dôme. At the summit the experiment was repeated, and the column now read 23 Paris inches, 2 lines. The difference was equivalent to about $3\frac{1}{2}$ English inches. The experiment at the top was repeated under various conditions—in the open air, under shelter, in a shower, in a mist, and always with the same result. On the way down it was repeated at successively lower levels, and the column showed a steady increase, until at the base it returned to its original value. This historic experiment established beyond all doubt the fact that the column of mercury was sustained entirely by the pressure of the atmosphere, and gave to the world the barometer as we now know it. It explained finally why the Italian pump could only draw water up the 40 feet pipe for a distance of 33 feet, in that this was the total equivalence in weight of the atmospheric pressure. Water could only be drawn by the pump to such a height as to make the total pressure inside equal to that outside, and although there remained 7 feet of pipe with no air, the water could not rise to fill this additional space because the air outside could force it no further. The doctrine of the ' horror vacui ' was no more.

5

IV. OTTO VON GUERICKE (1602–86)

For the next step in the evolution of the physics of
the atmosphere the scene now shifts to Germany, and
brings us to the activities of Otto von Guericke. Guericke was born at Magdeburg in Saxony in 1602. He
came of a family well known in the Saxon city of his
birth, and he received a good education, studying both
law and mathematics at various centres, including
the University of Leyden. He then spent some time
in travel in France and England, after which he returned
to Germany, and received the appointment of chief
engineer to the city of Elfurt. The Thirty Years' War
proved disastrous for his family, who barely escaped
from the devastations of Magdeburg with their lives,
but subsequently Guericke served as a military engineer.
In 1646 he was elected Burgomaster of his native city,
and it was during the leisure time of his thirty-five
years of service in this office that he devoted himself
to science in general, and in particular to those researches
on air-pressures that have since made him famous.
He relinquished his office in 1681 and retired to Ham-
burg, where he died in 1686.

Guericke was led to his experiments on air pressures
through an early interest in astronomy, and in particular
through an interest in the dispute as to what, if anything, lay in the space between the heavenly bodies.
Guericke argued that it could not contain air, as in
that case the friction would cause a slowing down of
the planets in their orbits. Hence he was led to the
problem of whether or no it was possible to create a
vacuum so as to create conditions comparable with
those in space. It is to his credit that, starting with
all the mediæval ideas current regarding the atmosphere
—that Nature abhors a vacuum, that air is not an
element but is the exhalation from bodies that we do
not perceive because we are accustomed to it from
childhood, and the like—he was able without hesitation

to abandon these views as the course of his experiments showed them to be untrue.

In his first attempt to produce a vacuum, Guericke employed a wooden cask filled with water, and applied a pump similar to the ordinary suction pump with the idea of extracting the water without allowing the air to enter in its place. It was very difficult to work the pump in consequence of the atmospheric pressure of 15 lb. per square inch, although he was in ignorance of this explanation. A hissing noise due to the admission of air through the faulty joints soon showed his failure. In spite of the repeated strengthening of the fastenings and the labours of three men on the pump, the air continued to pour in.

He now repeated the experiment with one cask inside another with the idea of interposing a water-cushion between the outer air and the interior of the inner cask ; but although pumping proceeded for several days, the inner cask, when ultimately deprived of water, was yet found full of air.

Guericke now came to the conclusion that he must employ a stronger vessel than one made of wood, and he therefore prepared a copper globe. The first few strokes of the piston were found easy, but soon the pumping became so difficult that the united efforts of four strong men were required. Yet in spite of this the sphere suddenly collapsed with a loud and alarming report—an experiment still carried out on a small scale in all school laboratories of to-day. His partial success made Guericke feel that all he needed was both a stouter globe and a more efficient pump, and when these at last became available, he repeated the experiment, and this time with complete success. His account of the re-admission of the air reads very impressively.

'On opening the stop-cock the air rushed with such a force into the copper globe, as if it wanted to drag to itself a person standing near by. Though you held your face at a considerable distance, your breath was taken away ; indeed you could not hold your hand over the stop-cock without danger that it would be violently forced down.'

We have seen that in every experiment so far Guericke had adopted the method of first filling the vessel to be evacuated with water. He now determined to attempt the evacuation of the air within directly, and for this purpose he devised the *air-pump*, an invention with which his name has ever since been associated. A large glass spherical receiver was mounted and fitted with a detachable stop-cock. Any object that it was desired to subject to reduced pressures was placed within the receiver by removing the mounting. To secure more perfect closure the receiver was connected with the pump in a vessel which was then filled with water. The pump was mounted on a tripod stand, and although its design was crude, it was the first of its kind, and it enabled Guericke to proceed at once to a series of interesting experiments of a fruitful kind.

The most picturesque and possibly the most famous was the experiment with the Magdeburg hemispheres. He constructed two exactly similar hemispheres of copper, each about 12 feet in diameter, and fitted them into an air-tight joint by means of a soft leather washer. The combined sphere was now fitted to the air pump and evacuated, with the result that, in the presence of the members of the Reichstag and the Emperor Ferdinand III, it required no less than sixteen horses, eight harnessed in each direction, to pull the spheres apart with a loud report that astonished the spectators.

Guericke performed a number of other spectacular experiments designed to illustrate the great pressure of the atmosphere on evacuated vessels, but there were also others, which, if they were not so spectacular, were more important. For instance, Guericke definitely proved that air has weight by weighing the receiver before and after the evacuation. Various properties of the vacuum were illustrated. The inability of sound to be transmitted except through a medium was shown by the introduction of a clock which was seen to work but whose striking was inaudible. The necessity of air to life was shown by the quick death of a bird or a fish introduced into the receiver. These and many other experiments were all made possible by Guericke, and it is little wonder that his memory is cherished at Magdeburg.

V. THE HON. ROBERT BOYLE (1627-91)

The scene for the final stage in the history of the physics of the air is set in England, where a work called *Mechanica hydraulico-pneumatica*, written by Kaspar Schott and published in 1657, was read by a famous English savant named Robert Boyle. This work included an account of Guericke's researches, and inspired Boyle to similar studies.

Robert Boyle was born at Lismore in Ireland in 1627. His father was the Earl of Cork, a man of ancient ancestry. As a child Boyle developed a stammer through the imitation of the stammering of some of his playmates, and this disability remained with him for life. He left Ireland for Eton at the age of eight, stayed there four years, and then continued his education with a private tutor, with whom, together with an elder brother, he went abroad in 1638. They journeyed leisurely through Holland, France and Switzerland and for a time settled at Geneva. Later, moving on to Italy, Boyle became acquainted with the writings of Galileo, and it is possible that these writings profoundly influenced him towards the study of science.

The death of his father brought Boyle home in 1644, and in spite of the distractions of the Civil War between Royalists and Parliament, in which he took no part, he confined himself to study and discussion with those of a like spirit. So began that 'invisible college' of gatherings which developed ultimately into the formation of the Royal Society, an event of undoubted importance in the history of science. It was an excellent example that was speedily followed in other countries. The original meetings in London dated from 1645, but by 1654 Boyle had definitely settled in Oxford, and

here he became absorbed with his many philosophical
inquiries, in collaboration largely with the well-known
Robert Hooke, whom he had engaged as his assistant.
For some time meetings were held both in London and
in Oxford, and it was in 1660 that King Charles II
granted the Royal Charter bringing to the Royal Society
formal recognition as the earliest and the premier scientific
society in England.

Boyle's life was one long devotion to science and
religion. He founded an organization which later
developed into the present-day Society for the Propagation of the Gospel in Foreign Parts, exhibited continual
evidences of charity, modestly declined all honours,
including both the Presidency of the Royal Society
and a peerage, was devoted to a sister, Lady Ranelagh,
with whom he lived (he never married), and died in
1691 aged sixty-four years. Essentially a member
of the experimental school, he confessed that ' in my
laboratory I find that water of Lethe which causes that
I forget everything but the joy of making experiments.'
He is rightly regarded as one of the outstanding personalities in the history of both physics and chemistry.

When Boyle was first attracted to the study of the
physical properties of the atmosphere, he quickly realized the necessity for a better and more efficient pump
than that which Guericke had devised. This he
accordingly produced in collaboration with Robert
Hooke, and it was such an improvement as to take
the form, in principle, of the air-pumps of to-day.
With this he at once carried out a number of important
experiments. His spherical receiver was so designed
as to enable him to introduce a small chemical balance
into its interior, suspended from the top, and by this
means he was able to carry out with greater accuracy
than before the experiment of finding the weight of a
sample of air.

One of Boyle's most effective experiments with his
pump was to apply the Torricellian experiment to the
case of artificially reduced pressures; in other words,

to show in a laboratory what before could only be shown by going up a mountain. The method was simply to suspend a Torricellian mercury column inside the receiver, and then to commence evacuation by working the pump. As the air was withdrawn, the mercury in the tube steadily fell, and Boyle was in fact able to get down to less than one inch of pressure. It was a striking laboratory verification of the true explanation of the cause of the existence of the liquid column.

Yet another important observation was the influence of pressure changes in the atmosphere upon the boiling point of water. He introduced into the receiver some water cooled to just below the temperature of its boiling point, and by working the pump the pressure was then reduced. Soon a point was reached at which the water again began to boil, showing that a reduction of pressure had the effect of lowering the boiling point.

We come finally to the great contribution to the physics of gases made by Boyle, with which his name has ever since been associated as Boyle's Law. He had become convinced that the general effect of the application of pressure to a given sample of air was to compress it, i.e., to diminish its volume, though the exact mathematical relationship between these two quantities—the pressure applied and the volume resulting—had not yet become a subject for his determination. In 1660 he published a work called *New Experiments touching the Spring of the Air*, in the course of which he tried to convey, by a sort of physical analogy, the effect of pressure on air. He pictured the air particles as behaving like spherical springs whose volume would diminish on compression. It was in consequence of some very absurd criticisms this book received that Boyle was prompted to those further experiments that led to what has since become known as Boyle's Law. He took a long glass tube bent as shown in the figure (Fig. 9(a)), the lower end being closed and graduated, and introduced mercury so that the levels were the same in the two limbs, and the volume of air imprisoned was measured in terms of its height. He now poured more mercury into the long limb until the air within was compressed to half its former volume, and found that the difference in the two columns was now 29 inches (Fig. 9(b)), so that the total pressure to which the im-prisoned air was subjected was two atmospheres, i.e., 29 inches of mercury plus the air pressure above (also equal to 29 inches). Readings were now varied for a number of mercury column values and the volumes of air corres-ponding were in each case obtained. Finally, by a suitable modification of his apparatus, using a very deep mercury reservoir, into which he lowered a *hot* tube corked at the top, he was able to extend his readings to the case of pres-sures below that of the atmosphere.

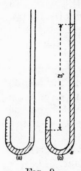

FIG. 9.
Establishment
of Boyle's Law

As the tube cooled the air within con-tracted and the mercury rose above that of the level in the reservoir, the difference in levels giving the excess of atmospheric pressure over that to which the sample of the air was subjected. In this way Boyle was able to deduce that the volume of the gas was inversely pro-portional to the pressure.

CHAPTER VI

HERALDS OF THE NEW PHYSICS

I. FRANCIS BACON (1561–1626)

THE examples of Galileo and Gilbert were not the only evidences of a new spirit in scientific inquiry. In the year 1561 there was born in Francis Bacon one who although not himself a man of science, was destined nevertheless to exert a profound influence on philosophy. Educated in Law at Trinity College, Cambridge, he entered Parliament at the age of 23, and soon afterwards published his ' Essays ' and became famous at one bound. He rose quickly, and by 1618 he became Lord Chancellor with the title of Lord Verulam. In 1620 he was accused of corrupt practices, pleaded guilty, and was removed from office shortly after the publication of his *Novum Organum* with its famous plea for the inductive method. The remaining few years of his life were devoted to literature and philosophy. It was the pursuit, indeed, of an investigation on the effects of cold in preventing putrefaction that brought him to his grave. He contracted a cold while stuffing a fowl with snow, and died after a week's illness, in 1626. Thus, as Macaulay remarks, the great apostle of experimental philosophy was destined to be its martyr. It is a coincidence that Bacon's last experiment should foreshadow the modern almost universal practice of meat preservation by cold storage.

Bacon's influence on the science of his day was very real, but has been the subject of considerable controversy. To him has been attributed the inception of the Inductive Method of investigation—the deduction from data derived by careful observation and experiment as to the causes of things. In fact, however, the principles of inductive philosophy had not only been practised before his time, but had been specifically formulated, though admittedly in a crude way. Lord Macaulay very amusingly illustrates the truth that the Inductive Method is in fact instinctive and everyday in its application.

'A plain man finds his stomach out of order. He never heard Lord Bacon's name ; but he proceeds in the strictest conformity with the rules laid down in the second work of the *Novum Organum,* and satisfies himself that minced pies have done the mischief. "I ate minced pies on Monday and Wednesday and I was kept awake by indigestion all night." This is the *comparentia ad intellectum instantiarum convenientium.* "I did not eat any on Tuesday and Friday, and I was quite well." This is the *comparentia instantiarum in proximo quæ natura data privantur.* "I ate sparingly of them on Sunday, and was slightly indisposed in the evening. But on Christmas Day I almost dined on them, and was so ill that I was in great danger." This is the *comparentia instantiarum secundum magis et minus.* "It cannot have been the brandy which I took with them, for I have drunk brandy daily for years without being the worse for it." This is the *rejectio naturarum.* Our invalid then proceeds to what is termed by Bacon the *Vindemiatio,* and pronounces that minced pies do not agree with him.'

Nevertheless, it is but just to say of Bacon that he was the first to analyse the inductive methods of reasoning with closeness and accuracy, and to bring into the form of a declared system principles which, though already adopted in practice, had never before so adequately been presented in their mutual relation and interdependence. Further, his thesis received such a wide publicity as to react on the public consciousness as never before, and accordingly produced for science (in which physics naturally shared) an impetus which certainly entitles him to our recognition in these pages.

From the point of view of our subject, Bacon, by applying his methods of reasoning, developed a theory of heat which, in the light of subsequent developments (see Chap. X) is well worthy of description here as an adequate illustration of the working of the Inductive Method.

His first step was to arrange lists or tables of all experiments and observations relating to heat. The first positive table was of those instances which ' agree in possessing the nature of heat, e.g., sun's rays, meteors, lightning, flame, hot springs, sparks, hay, slaked lime, the bodies of animals.' The second was a negative table parallel with the first, i.e., possessing similar qualities but devoid of heat. Thus parallel with the sun's rays which are luminous, are lunar rays, stellar rays, etc. In Table III (comparative) were comparisons of the degrees of heat found in different substances, from quicklime, green plants, etc., in which ' heat ' is insensible to the touch, and thence to degrees of heat in various flames of alcohol, wood, oil, sulphur, gunpowder, and lightning.

These tables are designed ' to present a view of instances to the understanding.' Now the process of *induction* is put into operation. ' Upon a particular and general view of all the instances, some quality or property is to be discovered, on which the nature of the thing in question depends, and which may continually be present or absent, and always increase and decrease with that nature.' By a process of elimination, the field of inquiry is contracted, the eliminants forming a fourth table. Thus as both sun's rays and a common fire are hot, Bacon excludes both the *terrestrial* and *celestial natures* as causes of heat. Similarly *luminosity* is excluded as being common to stars and moon, neither of which are hot. *Tenuity* is rejected because gold can be made hot, whereas air is generally cool ; and so on. Continuing, Bacon points out that ' the business of exclusion lays the foundations for a genuine induction, which, however, is not perfected until it terminates in the affirmative ; for a negative conclusion cannot possibly be perfect.' These affirmative conclusions constitute a fifth table, which Bacon speaks of as ' The First Vintage concerning the Form of Heat,' and his investigation ends :

' *Heat is an expansive bridled motion, struggling in the small particles of bodies.* But this expansion is modified, so that, while it spreads in circumference, it has a greater tendency upwards. It is also vigorous and active ; and as to practice, if in any natural body a motion can be excited which shall dilate or expand, and again recoil or turn back upon itself, so that the dilation shall not proceed equally, but partly prevail and partly be checked, any man may doubtless produce heat ; and this may serve as an example of our method of investigating Forms.'

It is obvious that Bacon's schemes of tables were cumbrous and clumsy. His omissions of the rôles of hypothesis and scientific vision show the clearest evidence of his own ignorance of science, its history and its possibilities. His rejection of the Copernican system and his belittling of Galileo and Gilbert almost rule him out of court. Nevertheless, the extravagant denials of his influence and his place in the history of science are as unfortunate to-day as were the super-abundant laudations of a century ago. His import-ance in the records of history must remain, as a man who, at a totally favourable moment, fanned into popular flame an interest in and a pursuit of scientific investigation without which progress can never be maintained.

II. RENÉ DESCARTES (1596–1650)

One other philosopher who aided in the impetus to science in the seventeenth century must now briefly be considered—René Descartes. He was born in 1596 in France, of wealthy parents. He received an indul-gent training as a child, and as a young man lived a life of ' quality.' In 1615, however, he came under the influence of Father Mersenne, a mathematician of some note, and began a course of serious study. In 1617 he took up a military career, but continued mean-while his devotion to mathematics and philosophy.

It was in 1619 that he developed, as the result of a sudden intuition, the beginnings of analytical geometry. He finally abandoned the profession of arms in 1621, and after an interval of travel he settled in Holland in 1628. Here the most fruitful years of his life were spent. In 1649 he joined the court of Queen Christina of Sweden as tutor, but a year later he contracted a chill and died.

Descartes, like Bacon, exerted a profound influence on the science of the seventeenth century. While, however, Bacon proposed the method of *inductive* reasoning founded on experiences of the *external* world, Descartes' path was by *deductive* reasoning from the *internal* consciousness. ' I think, and therefore I exist.' Descartes was therefore in a sense the complement of Bacon; and both rejected authority. Descartes' methods were essentially those of the mathematician. Starting with the enunciation of a definite principle, and working from it, he would proceed step by step from deduction to deduction until he developed a complete scheme. To him Nature's problems were as a tree of which metaphysics was the root and physics the trunk, with its three branches of mechanics (the external world), medicine (the human body), and morals (human conduct). All this was elaborated in his famous *Système du Monde,* completed in 1633 and published in 1644. It was a book that profoundly stirred Europe and the famous vortex theory that it contained received an astonishing degree of acceptance.

Descartes' physical writings cover a wide field in both optics and mechanics. He gives a complete account of the law of refractions, originally discovered by Snell. He makes no mention, however, of being aware of Snell's work on the subject. He also investigated a number of problems on the refraction produced by lenses, including the nature of the curved surface which must be given to the glass in order that parallel rays incident upon its surface may meet in an exact point. Descartes also investigated the explanation of the rainbow with considerable success.

To sum up the situation, we may say that by the accumulated influence of Galileo and Gilbert in the attack on dogma and in the inauguration of experimental science and of Bacon and Descartes in the enunciation of theories of scientific method, the impulse to progress became inevitably great in the seventeenth century. Amid the galaxy of men of science of the first rank who graced this century with their wonderful efforts, Sir Isaac Newton easily stands out as the greatest figure of all. To him and some of his activities we now proceed.

CHAPTER VII

NEWTONIAN PHYSICS

I. SIR ISAAC NEWTON (1642-1727)

SIR ISAAC NEWTON was born in 1642—the year of Galileo's death—at Woolsthorpe Manon, near Grantham. He came of yeoman stock of long lineage. At the age of twelve he went to the King's School, Grantham, as a day boy, staying at the house of a local apothecary. At this time he exhibited no special signs of genius other than the keen interest in the making of mechanical models. At the age of fifteen he left school. His mother intended him to take up farming, but by now he was thoroughly imbued with a desire to study, and he spent far more time at his books than he did on the land. His mother soon bowed to the inevitable— his father had died before Isaac was born—and back he went to the school at Grantham, this time with the specific purpose of entering Cambridge. He entered Trinity College, Cambridge, as an undergraduate in 1660. At once the genius in Newton began to blossom forth. Soon he was ahead of both lectures and text-books. He was fortunate, too, in coming under the influence of Dr. Isaac Barrow, the Lucasian Professor of Mathematics. It was the inspiration and friendship of this learned man that did so much for Newton, and among other things led directly to the invention by the young philosopher of the calculus—which, under the name of the theory of fluxions placed a new and most powerful weapon of mathematical investigation in the hands of science.

73

Newton took his B.A. in 1664. A year later the university was closed against an outbreak of plague, and so for the next two years we find him back at Woolsthorpe. This proved to be an interval of great importance, inasmuch as Newton during that time was occupied mainly on those speculations that led to his epoch-making enunciation of the famous inverse square law of universal gravitation.

In 1667 we find him back again at Cambridge, and the next few years, during the course of which he succeeded Dr. Barrow in the Lucasian Chair of Mathematics at the age of twenty-six, saw the completion of a large number of researches in physics, astronomy and mathematics of the most far-reaching character. Much of this work involved him in controversy with others, and in particular his theory of fluxions led to disputations as to priority against a rival claim by Leibnitz. The details, however, concern us but little in a work dealing only with physics, and we will therefore pass them over.

Newton's monumental work, the *Principia,* or *Mathematical Principles of Natural Philosophy,* was completed by 1687, and in spite of many obstacles, some of them financial and some personal, it was published through the good offices of the Royal Society. Except for a year's interlude in Parliament from 1688 to 1689, Newton continued steadily at his professorial duties, but in 1692 he suffered a serious breakdown in health. By 1694 he was fully recovered, however, and a year later he accepted the post of Warden of the Mint. He resigned his professorship finally in 1703.

Honours were conferred on him in abundance, but they in no way affected his modest shyness and innate sensitiveness. In 1699 he was enrolled as the first Foreign Associate of the Academy of Sciences at Paris, in 1703 he was elected President of the Royal Society, and in 1705 he was knighted by Queen Anne. Yet although his professorial days were over he remained in his old age a vigorous force in the world of learning. He never married, and lived with a niece. His later years were given up largely to a study of divinity, and were on the whole uneventful. He died on 20th March, 1727, at the advanced age of eighty-five years, honoured and mourned by the whole world, and nothing could be more impressive than the ungrudging tribute of his old academic enemy, Leibnitz, who said of him : ' Taking mathematicians from the beginning of the world to the time when Newton lived, what he had done was much the better half.'

And this is probably true of his work in science generally. No one is more conscious of his own shortcomings and ignorance than is the true man of science. Nature and her mysteries are so vast, and knowledge in its sum so relatively small, that we may well conclude this brief review of Newton's life by quoting his own famous estimate of his work.

' I know not what the world may think of my labours, but to myself it seems that I have been but as a child playing on the sea-shore ; now finding some prettier pebble or more beautiful shell than my companions, while the unbounded ocean of truth lay undiscovered before me.'

II. NEWTON'S PHYSICS—OPTICS

In Optics, a field Newton greatly enriched by his researches, his first work of note was on the subject of the analysis of white light, and it is interesting that the views on colour which were prevalent when he began his experiments were thus expressed by Dr. Isaac Barrow :

' White is that which discharges a copious light equally clear in every direction. Black is that which does not emit light at all, or which does it very sparingly. Red is that which emits a light more clear than usual, but interrupted by shady interstices. Blue is that which discharges a rarefied light, as in bodies which consist of white and black particles arranged alternately. Green is nearly allied to blue. Yellow is a mixture of much white and a little red ; and purple consists of a great deal of blue mixed with a small portion of red. The blue colour of the sea arises from the whiteness of the salt it contains, mixed with the blackness of the pure water in which the salt is dissolved ; and the

6

blueness of the shadows of bodies, seen at the same time by candle and daylight, arises from the whiteness of the paper mixed with the faint light of blackness of twilight.'

There is every probability that Newton was consulted by Barrow during the course of the formulation of his ' Theory of Colours ' in 1668, and we know that four years earlier, in 1664, Newton wrote to say that he had purchased a prism ' to try the celebrated phenomenon of colours.' Newton would certainly have put Barrow right in 1668 if his famous researches on the subject had been carried out by then. Actually, however, his results are referred to in a letter dated February 1669, so that the experiments must have taken place between

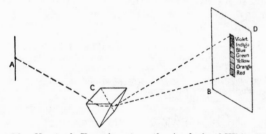

FIG. 10.—Newton's Experiment on the Analysis of White Light

1668 and that date. Colour trouble in refracting telescopes was the inspiration of all such experiments at that time. The telescope had become a weapon of research so full of potentialities, that clearly investigations with the object of trying to eradicate the fault we speak of as chromatic aberration had become an urgent necessity. The first step was to ascertain the cause of the trouble. The remedy would then inevitably follow.

Newton's experiment was to allow a beam of light A C to enter a darkened room through a round hole A (Fig. 10) and to fall on a prism C. The beam was refracted through, spreading out as it proceeded, and was intercepted on a white screen B D as a coloured band, about five times as long as it was broad, whose colour gradations always, beginning from the most refracted end, merged from violet to indigo, indigo to blue, blue to green, green to orange, and orange to red. Newton was puzzled by the great length of the spectrum in comparison with its breadth. From the laws of refraction, he expected a circular image instead of a band, and to seek the cause he first considered his apparatus. He tested successively whether the thickness of the glass, or the size of the aperture A, or veins and defects in the glass, could cause the elongation of the image, and was obliged to rule out all these considerations. Finally he decided to study each portion of the spectrum

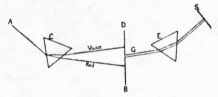

Fig. 11.—Passing on a Single-Coloured Ray through a Second Prism

separately. This he did by arranging for a pin-hole in the screen B D, so as to allow in turn each portion of the spectrum to pass on through the screen as a single coloured ray G. This ray was now made to pass through a *second* prism E (Fig. 11) behind the screen, and the extent to which this second prism refracted it was observed. Newton thus found that the violet light was more refracted than the blue, blue more than yellow, and so on. He thus showed that the lengthening of the image into a spectrum was due to the varying refrangibility, from red to violet, of rays which, in their aggregate, gave white light before they were spread out and separated by the prism.

Newton now proceeded to the reverse experiment of re-combining the coloured rays, one method being by bringing the rays to a common focus though a large lens. As he anticipated, he got white light (Fig. 12) as a result of this re-combination.

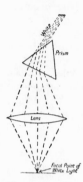

Newton realized now that inasmuch as a lens tends to behave like a complex system of prisms, the reason for chromatic aberration was that instead of getting one true focus for white light incident upon it, each colour of the spectrum had its own focal distance, the shortest being for violet light, and the longest for red light, giving a coloured region instead of a white focal point (Fig. 13). Moreover, he was under the impression that the spreading-out or dispersion of the spectrum was the same for all kinds of glass, and that chromatic aberration was therefore incurable (though we now know that for flint glass the dispersion bears a much greater ratio to the deviation than in the case of crown glass, and that this is the basis of the achromatic combination of lenses that practically cures chro-

FIG. 12. — Re-combination into White Light

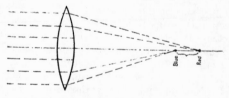

FIG. 13.—Cause of Chromatic Aberration

matic aberration to-day). He therefore decided to abandon the use of refracting telescopes, and turned his attention to the design of a *reflecting* telescope. This had been attempted, more or less scientifically, by James Gregory, but Newton's design (Fig. 14) in 1672 was undoubtedly superior. Rays R R from the object were reflected from a spherical mirror A B through a focus F, giving an image I_1, and this was then viewed by the observer through a small lens L in the *side* of the instrument through a small mirror M inclined at 45° to the axis of the telescope, giving a resultant image I_2.

Newton's experiments on colour were the subject of

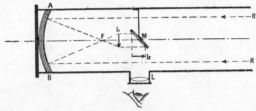

Fig. 14.—To illustrate the Principle of Newton's Telescope

violent attacks by a number of contemporaries, of whom the most vigorous was Robert Hooke, a man of marked ability, bad temper, and thwarted ambitions. Hooke criticized Newton's telescope and opposed Newton's views as to the composition of white light, which he held to be compounded of red and violet only. Hooke also submitted to the Royal Society certain observations on the colours produced by thin transparent films of soap bubbles and by very thin air layers between glass plates

Fig. 15.—To illustrate the Formation of Newton's Rings

nearly in contact. Hooke felt that the thickness of the film was somehow related to the effects produced, but it was left to Newton to solve the problem for him. Newton took two large lenses of known but differing curvature, one *a b* a convex lens (Fig. 15) of larger curvature than that of the concave surface *c d* to which it was applied. The effect he got is spoken of as *Newton's Rings*. The radius of each curved surface being known, it is easy to calculate any interval such as fg between the two from the point of contact e. The effect obtained was a series of concentric rings of various colours. If, however, light of only *one* colour is admitted to the combination, as by first passing the white light through red glass, a series of alternate light and dark rings are seen; and according to the portion of the spectrum from which the light is chosen, these ring systems are either close together, as with blue light, or further apart, as with red light, and so on. If the several series of rings obtained from each colour be superimposed on each other, the effect would be as when the film is viewed in white light. Working on these lines, Newton was able to find the precise thickness of the air-film corresponding to any simple colour.

The system of rings we have just discussed are those seen by reflected light from above. Newton saw, however, that when the light is allowed to pass through, another system of rings is seen, the reverse of the former, so that the centre, which was a dark spot, is now a light spot, and similarly for the bright spaces.

Newton expressed these facts by a theory that along a ray of light there were regularly alternating positions at which the ray had what he called *fits of easy reflection* and *fits of easy transmission*. Newton's measurements were remarkably accurate, and his results still stand good, although his theory has proved inadmissible.

Newton was able to explain the various observed facts of optics fairly easily by the famous emission or corpuscular theory of light. According to this a luminous body emits streams of minute corpuscles that move in straight lines, and the sense of vision is produced by the impact of these streams on the retina of the eye. On this theory refraction is accounted for by the assumption that as each corpuscle of a stream approaches a denser refracting surface, i.e., the surface of denser medium at any given angle, it begins to be attracted towards it, so that the component of the velocity of the corpuscle along the normal is increased—or in the case of a rarer medium the normal component velocity is diminished—whilst in both cases the component of the velocity perpendicular to the normal remains unaltered. The resulting ray thus agrees in the bending towards or away from the normal with the results of experiment, and the velocity of the particle becomes faster through the denser medium, and slower through the rarer medium. The more difficult point, that in a transparent medium *some* of the rays are reflected back and some refracted through at the same time, was explained as we have pointed out above by Newton's view of the existence along the ray of alternate ' fits of easy reflection and easy transmission.' Finally it should be noted that this corpuscular theory definitely assumes the existence of an ' ether ' through which the corpuscle streams must pass.

III. HUYGHENS AND HIS UNDULATORY THEORY

The rival to the emission theory of light, thus supported by Newton (although he was not absolutely committed to it), was the undulatory theory of Christian Huyghens. Huyghens was one of the most illustrious of the men of science of the seventeenth century. He was born at The Hague in 1629, and was intended for the legal profession. He studied law at the University of Leyden. Mathematics soon attracted him, however, and in 1665 he was invited to Paris by Louis XIV. His stay in Paris lasted from 1666 to 1681, after which he returned finally to Holland. He died in 1695, after a life devoted to the enrichment of science in astronomy, mathematics and physics. Among his many inventions, none is more noteworthy than his application of the principle of isochronism of the pendulum, originally discovered by Galileo, to the well-known arrangement by which the oscillations of the pendulum are made to control the motion of the clock. He proved that for complete isochronism the curve in which the suspended body must move is not an arc of a circle, but a cycloid, and he worked out the mathematical consequences of this in a most brilliant fashion.

The undulatory theory of light was first put forward by Huyghens in a communication addressed to the Academy of Sciences at Paris in 1678, and was elaborated in his *Tracté de la Lumière,* published in 1690. A similar theory had also been put more crudely by Robert Hooke in 1665. According to this theory, all space is pervaded by an inconceivably subtle and elastic medium, or *ether,* through which waves or undulations are propagated from a light-source in all directions. When these undulations reach the eye the sensation of sight

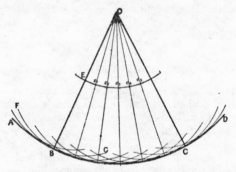

Fɪɢ. 16.—Huyghens' Theory of Secondary Waves

results. Under ordinary circumstances, the undulations spread from the point of origin in a regular spherical form, just as the waves produced by dropping a stone into still water spread out in circles. Thus in the figure (Fig. 16), if O is the light source, let A B C D be a portion of the spherical wave transmitted through the ether. Then any particle E within this wave will, as soon as the original wave has reached it, itself become the centre of a new disturbance, setting up a wave F B G which touches the original wave A B C D at B. Similarly every other point within the sphere sets up similar secondary waves, and it will be seen that all the spheres like F B G produced from points e_1, e_2, e_3, etc., and equidistant from O, touch the sphere A B C D ; indeed it is this ' envelope ' of these secondary wavelets that produces the ' wave-front ' A B C D.

This theory was applied by Huyghens to the explanation of all the ordinary phenomena of reflection and refraction. For refraction it required the assumption that *the velocity of propagation is less in the denser medium than in the rarer medium,* as a result of which a change of direction of the wave-front is produced. In the figure (Fig. 17), let A B be the front of the wave advancing in the direction of the arrow. The portions of the wave successively entering the denser medium will move on with decreased velocity, and while the part of the wave at B is still advancing in the rarer medium, A must have moved in the denser medium to a point A^1 where A A^1 is less than B B^1.

Suppose, for example, the speed in the denser medium

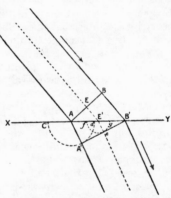

FIG. 17.—Explanation of Refraction in terms of the Wave Theory

is two-thirds of that in the rarer medium, then, if we describe a circle C A^1 d with centre A and radius equal to two-thirds of B B^1, when B arrives at B^1, the point A must have reached the circle C A^1 d. Now consider the point E in the wave half-way between A and B ; it reaches E^1 in half the time required for B to reach B^1, and the other half of the time it will be moving in the denser medium, reaching a point e in the circle $f e g$ described with centre E^1 and a radius of two-thirds of E E^1. Hence the wave front when B arrives at B^1 will

be the straight line A^1 e B^1 which passes through B^1
and touches the two circles ; for it will be evident that
whatever point along A B is considered, its place will
be determined by a proportionate circle to which A^1 B^1
is a common tangent.

In 1669 Erasmus Bartholinus of Copenhagen observed
that an object viewed through Iceland spar showed two
images instead of one—a phenomenon known as double
refraction, the incident ray being split up by the denser
medium into differently refracted rays. One of these
obeys the ordinary law of refraction and is known as
the *ordinary ray*—the other is the *extraordinary* ray.
Huyghens was able to explain this phenomenon on his
theory by assuming that the undulations in Iceland spar
corresponding to the extraordinary ray are propagated
in a *spheroidal form* at the same time as the ordinary
ray is propagated in the spherical form.

Such then were the two rival theories of light of the
seventeenth century—the emission theory and the
undulatory theory, each supported by a giant personality
of science. Controversy raged freely round the rival
theories, but that of Newton prevailed through the sheer
majesty of his major brilliance. He was in fact wrong
and Huyghens was right, but unfortunately the one
crucial test—an experiment that would show whether
the speed of light was greater or less in a denser medium
than in a rarer medium—was beyond the powers of the
day, and so the wrong view prevailed.

IV. NEWTON'S PHYSICS—MECHANICS

We turn finally to Newton's work in Mechanics.
Here his crowning achievement was undoubtedly his
discovery of universal gravitation and the formulation
of the famous inverse square law. It was beyond question the greatest gift ever offered to civilization by one
man. As an intellectual performance combined with an
exhibition of vivid imagination and intuition it is without
equal in the history of science. How did he reach his
results ? On the one hand, celestial mechanics was for

him like a natural magnet.   The problem of the motions
of the planets fascinated him.   Copernicus had enunciated to the world the hypothesis of a solar system of
planets moving in circles, and Kepler had elucidated
from the observations of Tycho Brahe the true celestial
framework of this system in his three famous laws of
planetary motion, viz., (1) that the planetary orbits are
ellipses with the sun at the focus, (2) that the planets
sweep equal areas to the sun in equal intervals of time,
and (3) that for any one planet the distance (d) from
the sun was related to the periodic time of revolution
(t) by the formula $\dfrac{d^3}{t^2} =$ constant.   These laws provided
the *how* of the framework ;   Newton sought the *why*.

   On the other hand, Newton was intrigued, as we know
from the historic anecdote of the falling apple, with the
reason *why* bodies fall to the ground when free from
restraint.   Here were apparently two separate problems
—the celestial one and the terrestrial one.   It was the
genius of Newton's imagination that linked and welded
them into one.

   Now Huyghens had already discussed with clarity
the problem of motion in a circle.   He had shown that
for a body to move in a circular path, it must be continually deflected from a rectilinear path by a constant
force directed towards the centre and with an acceleration equal to $\dfrac{v^2}{r}$, where v is the velocity of the body in
the circular path, and r the radius of the circle.   The
planets and satellites—for Newton was more immediately
concerned with the motion of the moon round the earth
—were moving in approximate circles.   Here was a
link of some significance.   The moon was held to its
path round the earth by an attraction proportional to an
acceleration of $\dfrac{v^2}{r}$;   a falling body was also attracted to
the earth.   If the attraction of the earth functioned on
a body not only on the surface of the earth, but also

deep in a mine and high up a mountain, why should there be any upper limit ? Why then should it not persist at the distance of the moon.

So Newton came to realize that the same acceleration that is possessed by a falling stone also prevents the moon from moving away in a rectangular path, whilst on the other hand its tangential velocity keeps it from falling towards the earth. He now turned to Kepler's laws, and at once found that the Law of Equal Areas was an immediate mathematical consequence of the assumption of this constant planetary acceleration towards the sun. The third law, that $\dfrac{d^3}{t^2} =$ constant, was now able to supply the nature of this central acceleration. For assuming for simplicity that the orbits are circular, the acceleration $= \dfrac{v^2}{d}$, where $v =$ velocity of the planet and $d =$ radius of the orbit.

Now $v = wd$, where w is the angular velocity, and $w = \dfrac{2\pi}{t}$.

Hence $v = \dfrac{2\pi d}{t}$, and the acceleration becomes

$$\frac{v^2}{d} = \frac{4\pi^2 d^2}{dt^2} = \frac{4\pi^2 d}{t^2} = Q.$$

Hence $t^2 = \dfrac{4\pi^2 d}{Q}$, and substituting in Kepler's Law III

$$\left(K = \frac{d^3}{t^2}\right)$$

we get $\qquad K = \dfrac{d^3}{\dfrac{4\pi^2 d}{Q}} = \dfrac{d^3 Q}{4\pi^2 d} = \dfrac{d^2 Q}{4\pi^2}$

whence $\qquad Q = \dfrac{4\pi^2 K}{d^2} = \dfrac{\text{a constant}}{d^2}$

i.e., the acceleration towards the centre varies inversely on the square of the distance.

Such was the nature of Newton's famous discovery, and it has been the pivotal law at the basis of what has been known as Newtonian Physics ever since.

Immediate consequences to formal mechanics were now seen. The fact that the same body suffers varying accelerations according to its position in space indicated the idea of a *variable weight* and a *constant mass*. The distinction between mass and weight thus emerged.

Newton successfully followed up the work of Galileo and Huyghens, in enunciating his well-known Three Laws of Motion :—

' Law I : Every body perseveres in its state of rest or of uniform motion in a straight line, except in so far as it is compelled to change that state by impressed forces.

' Law II : Change of motion (i.e., of momentum) is proportional to the moving force impressed, and takes place in the direction of the straight line in which each force is impressed.

' Law III : Reaction is always equal and opposite to action ; that is to say, the actions of two bodies upon each other are always equal and directly opposite.'

Preceding his enunciation of these laws were a number of definitions, including that of force, and in fact Laws I and II are linked up with that definition. Following the laws were also a number of corollaries, two of which deal with the principle of the parallelogram of forces, which Newton was the first specifically to formulate, and a third dealt with the quantity of motion generated in the mutual action of bodies, a fourth with the fact that the motion of the centre of gravity is not changed by the mutual action of bodies, and two others dealt with relative motion.

Law III on the equality of action and reaction was an undoubted contribution of importance in physics. Galileo's principles alone were insufficient to solve problems concerning the motions of bodies that exert a mutual influence upon each other. This was recognized by Huyghens in his investigations on the centres of oscillation, but it was left to Newton to formulate the definite law of reactions. A body that presses or pulls another body is pressed or pulled equally by that other body. We will not elaborate further these important researches. Faulty as they are now known to be in certain particulars, they so completely revolutionized mechanics as to place this subject at a bound in the forefront of modern science. Newtonian Physics has dominated the world of science ever since, and its founder has held undisputed sway comparable only with that of Aristotle in the days before him. The challenge to the ultimate permanence of Newtonian Physics which has recently appeared through the researches of Einstein in no way diminishes the glory of the name of Newton. He must go down in the records of history for ever as one of the greatest contributors to the advancement of civilization.

CHAPTER VIII

THE FOUNDERS OF THE QUANTITATIVE STUDY OF HEAT

I. THE PHYSICISTS OF THE THERMOMETER

THE eighteenth century proved to be an important one for physics. The modern outlook had become definitely established. The method of the laboratory was now a matter of routine. Measuring instruments were attaining a degree of efficiency that made for exactness of observation. The tide of investigation accordingly set in, with a flow that opened up all branches of physics, and with a speed that swept away all old standards and brought into being as recognized branches of the subject what before had been too early in growth to achieve a separate existence. Heat, light, sound, magnetism and electricity were now coming into their own as natural phenomena meriting individual and separate attention.

It is to the development of the subject of heat as a quantitative study that we propose to devote ourselves in this chapter. Heat as a separate study was late in development. At the beginning of the eighteenth century it was yet in its infancy while optics had pursued a vigorous growth from early times. There were still many people who regarded heat and cold as two distinct opposites. The reason for this late start was simple. No study can reach its true scientific stage until its concepts are capable of measurement. Until the instruments of measurement become available, little is possible beyond mere conjecture and often wild speculation.

The first instrument of measurement connected with the study of heat was the thermometer, and this only became available, and then as we have seen only in a very crude form, in the time of Galileo. At best it only indicated changes of temperature. It supplied no standards of comparisons. The first advance in this respect comes in fact from Newton, who in 1701 described in the ' Philosophical Transactions' an instrument with fixed points as standards of reference. Newton employed linseed oil as his expansible liquid, and for his lower fixed point he immersed the thermometer in snow, and marked the point at which the oil congealed. He chose as his upper fixed point the temperature of the human body. The interval between the two was divided into twelve degrees, and the divisions were extended upwards above the upper fixed point. On this instrument Newton found water to boil at 34°, tin to melt at 72°, and so on.

The disadvantage of Newton's instrument was due to the fact that linseed oil expands very little with heat. The instrument was therefore not sensitive. To remedy this, Guillaume Amontons (1663–1705) in 1702 returned to air as his thermometric substance. It was in principle the thermometer of Galileo and his Florentine colleagues, except that the barometer and its effect on the instrument being now known, corrections had to be applied to allow for changes in atmospheric pressure. Amontons, who suffered from deafness and regarded it as a blessing in securing him against interruption during his experiments, had found that the temperature of boiling water was constant, and he was the first to employ this temperature as his upper fixed point. His thermometer, nevertheless, found little favour because, owing to the use of air as the expanding substance, the instrument was necessarily big and cumbrous.

The first instrument to find general favour was the design of a scientific instrument-maker of Dantzig named Gabriel Fahrenheit (1686–1736). He had received a business education at Amsterdam, but a later interest in physics prompted him to travel in England and Europe. He was elected a Fellow of the Royal Society of London in 1724, the same year in which he published an account of his thermometer. He was stimulated to his researches by the work of Amontons, and he was in particular greatly impressed by the latter's observations on the constancy of the boiling point of water. This prompted him to test other liquids for their boiling points, and he was delighted to find that in each case there was a constant value appropriate to each liquid. He also noticed the influence of barometric changes on the boiling point—a fact we have seen to have been noted also by Boyle. Fahrenheit at first used alcohol as his thermometric liquid, but he soon turned to mercury as being in every way more suitable. His lower fixed point was found by plunging the instrument in a mixture of sal-ammoniac and snow as giving what he thought the lowest temperature he could obtain. This was marked 0°. The upper fixed point was that of boiling water, and was marked 212°. The figure was probably chosen because the temperatures of ice and of the human body came out at 32 and 96 respectively, but this is merely conjecture. The thermometer at once came into general use in England and Germany, and has remained popular ever since, although a rival system, adopted by Réaumur (1683–1757), a French physicist unfamiliar with Fahrenheit's work, found and still finds great favour in Russia and France. Réaumur used alcohol, and called the melting point of ice 0°, and the boiling point of water 80°.

Finally, in 1741 Andreas Celsius (1701–44) a professor of astronomy at Upsala, proposed a thermometer marked from 0° to 100°, though curiously enough he called the lower part 100° and the upper part 0°, and this scheme, reversed to its present form about 1749, has since been universally adopted for scientific purposes.

7

II. JOSEPH BLACK (1728-99)

The advent of the thermometer at once provided the means for discovering the true laws of heat, and the pioneer of this new advance was Joseph Black. This man of science, famous equally as chemist and physicist, was born at Bordeaux in 1728. His father, who had settled there in business, was a native of Belfast, and of Scottish descent. At the age of twelve Black was sent to a school in Belfast, and thence proceeded four years later to the University of Glasgow. Although strongly drawn to chemistry, he had chosen a career in the medical profession, and he completed his training at Edinburgh for this purpose. His stay at Edinburgh was associated with a number of masterly investigations in chemistry, and in 1756 he returned to Glasgow to occupy a chair in anatomy and chemistry, and later in medicine. He also developed a lucrative private practice, and rapidly acquired a name both for his professional and his professorial abilities. He died in his seventy-first year.

It was during the three years between 1759 and 1762 that Black completed those important researches on heat that brought the subject to the rank of a quantitative science. Their results were more far-reaching than this, however, for in the hands of James Watt they aided very materially in the invention and perfection of the steam engine, with lasting benefit to mankind. Let us inquire briefly into their nature.

The thermometer measures not quantity of heat, but differences in temperature or intensity of heat. Independently of any theories of heat it was obvious to Black that to raise the temperature of 2 lb. of water from 60° to 70° requires twice the *quantity* of heat that would be required in the case of 1 lb. of water. Here then was the first notion of quantity in relation to heat. Black showed as a result of a number of experiments that the quantities of heat required to raise the temperature of a given weight of water are proportional to the number of degrees through which the temperature is raised, and also that the temperature of a mixture of equal weights of water at different temperatures is equal to their mean. When however he mixed say 1 lb. of water at 150° with 1 lb. of turpentine at 50°, instead of getting the mean (i.e., 100°), the resulting temperature was 120°. In other words, the heat lost by the water would raise the same quantity of turpentine through a much greater range of temperature than for water, since we see that the quantity of heat which would raise 1 lb. of water 50° would raise 1 lb. of turpentine through 70°. The difference in the effects produced by the same quantity of heat is still more marked in the case of mercury and water. Thus if 1 lb. of mercury at 66° is mixed with 1 lb. of water at 100°, the resulting temperature is 99°, i.e., the quantity of heat that raises 1 lb. of water from 99° to 100° raises 1 lb. of mercury from 66° to 99°. In other words it requires 33 times as much heat to raise 1 lb. of water 1° as is required by 1 lb. of mercury. Black expressed this by using the term *capacity for heat* for different bodies, and it is a term that is independent of any views as to the nature of heat itself. This point is both interesting and important, because at that time the rise in temperature noted in bodies when subjected to pressure was considered to be due to a decreased capacity for heat, as though heat were a subtle fluid filling the pores of bodies, and was therefore squeezed out by pressure.

The *specific capacity for heat* was defined as the quantity of heat required to raise the temperature of 1 lb. of a substance 1° as compared with the quantity for a similar effect on 1 lb. of water, and it became contracted to the term *specific heat*. Values for the specific heat of various substances were worked out by friends and colleagues of Black, all based on the method of mixtures already alluded to.

We now turn to the other important concept developed by Black—that of *latent heat*. When a liquid, say water, is heated, on reaching 212° F., it begins to boil and the temperature remains fixed at 212° F. Also it takes a considerable time for the water to boil away. Again, when ice at 32° F. is heated and begins to melt, the temperature remains constant and does not rise until all the ice disappears. What, asked Black, becomes of the heat which all along has been given to the ice, but which has failed to raise its temperature ? He decided that the heat was in some way being utilized in the conversion of the ice into water, and that although not indicated by a change of temperature, it nevertheless existed in a *latent* form. This *latent heat* was therefore capable of measurement, and was the amount of heat required to convert 1 lb. of ice at 32° into water at the

Fig. 18.—Black's Ice Calorimeter Experiment

same temperature. It worked out to 143 heat units. Black spoke of it as the ' latent heat of fusion of ice.' Similarly there is a value for the latent heat of vaporization of steam, and indeed there are, as Black found, similar values for other substances. Thus he found that 1 lb. of solid tin is changed into 1 lb. of liquid tin without change of temperature by the absorption of about 26 units of heat. The value he found for the latent heat of steam was 990 units of heat.

Black employed these values to find the specific heat of a substance with very great accuracy, using what has since been called Black's ice calorimeter (Fig. 18). He procured a block of ice, ground one face flat, and holed out a cavity. A lid was formed of another block, so that it would be impossible for heat to reach the cavity from without. The body whose specific heat was required was first weighed, then heated to a known high temperature, and then quickly introduced into the cavity (which had been previously dried with an ice-cold cloth). The lid was then put into position. After some hours the body had cooled down to 32°, having given up its heat to the ice in the cavity, and for every 143 heat units lost by the body, 1 lb. of ice in the cavity was melted into water. This water was quickly wiped up and absorbed by a previously weighed piece of dry linen, which was now re-weighed, thus giving the total heat lost by the hot body. A simple calculation then sufficed to give the specific heat of the body.

These, then, were the discoveries of Black. They were the beginnings of modern calorimetry, and they made possible heat measurements whose results have contributed some of the most important data of science.

CHAPTER IX

THE INAUGURATORS OF ELECTRICAL SCIENCE

I. FOUNDERS OF STATICAL ELECTRICITY

ASSOCIATED with various natural phenomena, there emerged by the beginning of the eighteenth century a doctrine of imponderable and invisible fluids. Thus we have seen that there was a corpuscular theory of light; the imponderable fluid held to be squeezed out of a body when heat is applied was known as *caloric*; and *phlogiston*, another imponderable, was considered to be extracted from bodies during the process of burning. It is not surprising, therefore, to find that a body charged with electricity was regarded as emitting an *effluvium*. There was much that was vague in all this. The idea of imponderables was an impression rather than a conviction. But at least it was the starting point from which the experiments of the eighteenth century on electrical phenomena developed.

The first experiments of note were by Stephen Gray (1695 (?)–1736), an Englishman, who investigated the varying extent to which substances could conduct electricity from one charged body to another. This transference of ' electric virtue ' from one body to another was a new fact, and lent colour to the idea of the existence of an electric fluid. Dr. J. T. Desaguliers, to whom Gray had communicated the details of his experiments, distinguished two main classes—electrics and non-electrics, corresponding to the present-day distinction between non-conductors and conductors.

Meanwhile a specific electrical theory was being put forward by Charles François du Fay (1698–1739). Originally a French military officer, he quitted the army and devoted himself to science. He was an active member of the French Academy of Sciences, and succeeded the famous Buffon as Superintendent of the Royal Botanical Gardens at Paris. He was inspired to his electrical studies by the writings of Gray. He found that *all* bodies, including those that were apparently 'non-electric,' were capable of electrification by friction with a dry warm cloth, and that those which had previously been regarded as 'non-electrics' were in fact such good conductors of electricity that they tended to lose their charges as fast as they were received, to any suitable substances in contact with them, e.g., to the human body. Du Fay distinguished two kinds of electricity—'vitreous' and 'resinous,' because while any of the first group of charged substances (e.g., glass, rock-crystal, precious stones, hair of animals, wool, etc.) repels any other member of the same group, it attracts a charged number of the second group, (e.g., amber, copal, gum-lac, silk, thread, paper, etc.). This opened up distinct possibilities of development.

Towards the middle of the eighteenth century, another electrical phenomenon was discovered—that of the *Leyden Jar* by Musschenbroeck (1692–1761) at Leyden in 1746. We will describe it in his own words.

'I wish to describe to you a new but dangerous experiment which I advise you not to attempt yourself. I was engaged in some researches on the power of electricity, and for that purpose I had suspended, by two blue silk lines, a gun-barrel, which received the electricity of a glass globe rapidly turned on its axis and rubbed by applying the hand to it. At the end of the gun-barrel, away from the globe, there hung a brass wire, the end of which plunged into a round glass bottle, partly filled with water. I was holding the bottle with one hand, and with the other I was trying to draw sparks from the gun-barrel, when suddenly the hand holding the bottle was struck with so much violence that my frame was shaken as if by a lightning stroke. I thought that all was over with me, for my arms and my whole body were affected in a dreadful way, which I cannot describe.'

Musschenbroeck added that 'he would not take another shock for the Kingdom of France.' The electric shock, in spite of his alarm, nevertheless soon became the fashionable rage, and electrical experiments received an impetus in consequence, and William Watson (1715–87) suggested as an explanation of the Leyden Jar that it afforded a means for the storage of electricity at the expense of some other body, e.g., the friction machine in Musschenbroeck's original experiment.

A great contribution to the subject was now to come from an American citizen, Benjamin Franklin. He was born in 1706 at Boston, and was the youngest of a family of seventeen. His father was a soap-maker, and at the age of ten Benjamin was helping him in the business. Later he joined a brother as a printer's apprentice, and speedily became an avid reader of books. Some years of wandering followed, and after eighteen months' stay in London he returned to America, and at twenty-one opened an establishment of his own at Philadelphia as printer and stationer. He rapidly rose in the public esteem, and was elected to many public offices. The electrical experiments which were exciting all Europe caught his interest, and he procured some apparatus of his own from England. Then followed the researches that have made him famous, and that he described in 1747 in letters to Peter Collinson, a friend in England who was a Fellow of the Royal Society.

Franklin drew attention to the 'wonderful effect of pointed bodies, both in drawing off and throwing off the electrical fire.' 'We do not know whether this property be in lightning; but since they agree in all the particulars on which we can already compare them, it is not improbable that they agree likewise in this. Let the experiment be made.' The sight of a boy flying a kite suggested a method of readily and easily raising a metallic point towards the clouds, and he constructed a kite to which he attached a pointed wire. At the next approach of a thunderstorm he raised the kite, and succeeded, by the time the string was damp, in drawing sparks to his knuckle from a key at the end of the string. The identity of lightning with electricity was thus clearly established and lightning conductors followed as a practical application. It is interesting to note the controversy that arose in this connection as between the advocates of the use of ball-ended and of point-ended lightning conductors respectively. It appears that King George III had knobs put upon the conductors in the palace, and that this was commented upon in the following epigram :

> While you, great George, for knowledge hunt,
> And sharp conductors change for blunt,
>   The nation's out of joint,
> Franklin a wiser course pursues,
> And all your thunder useless views,
>   By keeping to the point.

Franklin's other great contribution was the substitution of a *one-fluid* theory of electricity for the two-fluid theory previously advanced by du Fay. He considered that every body in nature contained normally some definite quantity of a subtle fluid that constitutes electricity, and being normal, no signs of its existence are manifested. But when a glass tube is rubbed, it receives *more* than its normal share at the expense of the hand that rubs the tube. If, however, the person rubbing is standing on the ground, this deficiency is at once supplied from the earth, and he manifests no electrical signs whatever. Franklin expressed the conditions of ' more than normal ' and ' less than normal ' by the terms *positive and negative electrification*.

Franklin received many honours, both scientific and civic, during his lifetime, and he died in 1790 at the age of eighty-five years.

The whole study of electricity had now reached a stage at which little further progress could be made without measurement and quantitative research. These were happily now forthcoming from the hands of Charles Augustin Coulomb (1736–1806). He was a native of Angoulême, and was educated at Paris. At an early age he showed an extraordinary ability in the mathematical sciences, and at first took up the profession of military engineering. He served with the French army in the West Indies for some years, and then returned to Paris. Here he engaged in a number of investigations involving elaborate measurements and calculations, and was a member of the commission for the determination of new weights and measures decreed by the Government of the French Revolution which brought into being the metric system. It is related of him that when he was offered compensation for wrongful imprisonment by the government of Bretagne, all he asked for was a seconds

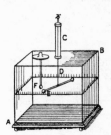

FIG. 19.—Coulomb's Torsion Balance

watch which he could use in experiment, and the famous Thomas Young said of him, ' his moral character is said to have been as correct as his mathematical investigations.'

Coulomb realized the necessity for a means of measuring the forces of electrical attraction and repulsion, and in 1777 he designed the *torsion balance* for this purpose. It was an ingenious and delicate instrument adequately sensitive to the very small forces it was designed to record. The upper plate of a cubical case of glass A B (Fig. 19) was pierced with two openings. Fixed into the central one was a glass tube C at the top of which was a screw head graduated in degrees. From this was suspended a hair fibre D and a gilded pith ball E. Into the other opening was inserted a fixed pith ball F, also gilded. The force of repulsion between the two swung E round from F, and the screwhead was then rotated to bring E back to its original position. The angle through which it was rotated was then a measure of the repulsive force. By this means Coulomb was able to show that forces of both electrical and magnetic attractions and repulsions varied inversely as the square of the distances between the charges. He also proved that electric charges exist only on the surfaces of conductors, and he was able to compare the surface charges on the various parts of the same conductor.

Mention should at the same time be made of the important researches of the Hon. Henry Cavendish (1731–1810). These extended over a very wide range, and were fully quantitative. He was the first to utilize the Leyden Jar in the form of the present-day condensers, whose capacity he was able to measure; he brought into being the term *potential*; and he compared the conductivities of many substances.

II. PIONEERS OF CURRENT ELECTRICITY

So far we have met with no knowledge of the electric current as such. All the experiments were directed to a study of electric charges and electrified bodies. The first steps in the recognition of an electric current carry us towards the end of the eighteenth century, when Luigi Galvani (1737–98), a professor of anatomy at Bologna, happened to leave a dissected frog on the table near an electrical machine. . An assistant by chance touched a nerve in the frog's leg with a scalpel, whereat the leg gave a kick. Galvani at once attributed this to the presence of the electrical machine, and decided to test the effect of lightning. He accordingly fastened a number of frogs' legs by brass hooks to an iron lattice outside his house, and certainly noticed kicks, particularly during storms. Of greater importance, however, was his observation that the legs kicked whenever he pressed the brass hooks against the iron lattice. Finally, working in his laboratory again, he produced violent kicks by placing the legs on an iron plate and then pressing the brass hook firmly against it. His conclusion was that *animal electricity*, or *galvanism*, manifested itself when two different metals were in contact with the nerves and muscles.

A compatriot of Galvani's, Alexander Volta (1745–

1827), now came forward to challenge his theory. He had noticed that a single combination of two metals held against the tongue produced a bitter taste if they were joined by a wire, and he expressed the opinion that the nerve was not a necessary factor in the production of the electricity. Volta was a professor of natural philosophy at Pavia for twenty-five years, from 1779 to 1804. He had long been experimenting on electricity, and in a letter to the Royal Society of London in 1800 he described an important discovery. Two plates of different metals (one always zinc, and the other either copper or silver) were placed in contact one over the other. Over this he then placed a piece of blotting paper or leather moistened with water or brine. Then followed another pair of the metals, then again blotting paper, and so on, until he had built up some two or three dozens of each kind of disc. This combination, he found, at once multiplied considerably the effect of a single set. It became known as the *Voltaic Pile*. Volta also found that when the discs were moistened with saline or acid solution, the effects were far more powerful than when water only was used. Volta found that the pile rapidly diminished in power when in use, and he attributed this to the drying up of the moisture in the porous discs. Accordingly he modified his apparatus to form a ' crown of cups.' Each of a series of vessels contained a saline solution into which was dipped a plate of zinc and a plate of copper. The copper of one vessel was joined by a wire to the zinc of the next, and so on to form a continuous chain. Such a combination gave an apparently inexhaustible supply of electricity.

Thus was given to the world the voltaic cell, the fore-runner of all modern batteries, and the beginnings of the study of current electricity and the electric circuit. The contact theory upon which Volta based his explanation of the cell's action did not survive long. It was soon to be replaced by a chemical theory, which found a justification in observations on the dissociation of the liquid elements in the cell, and on the practical and commercial consequences of electrolytic action. However, Volta had played his part. As in so many other cases of a young and growing branch of investigation, the study of the electrical circuit now needed to be placed on a quantitative basis. The need was supplied by Georg Simon Ohm, to a brief consideration of whose work we will now pass.

Ohm was born at Erlangen in Germany in 1789, and was the elder son of a locksmith who, having himself had the benefit of some education, made considerable sacrifices in order to send his sons to the university of his native town. They were only able to complete three terms, however, and Simon took a post as private tutor until he was able, by 1811, to save up sufficient to return and complete his graduation. He now settled in Bamberg as a teacher but owing to the sudden closing of the school at which he taught he was obliged for a time to return and help his father as a locksmith. This was bitterly disappointing to him, but he persevered with his studies, and in 1817 was again able to take up a post as teacher at the Jesuit High School at Cologne. It was during the nine years of his stay here that he carried out those researches on the laws of the electric current that subsequently brought him fame. The experiments were performed with home-made apparatus, and in 1827 he published his results in a work entitled *The Galvanic Chain, Mathematically Worked Out*. Curiously enough, it was badly received. Ignored by many, abused by some, Ohm finally became the victim of the hostility of the Minister of Education, and he resigned his post, and for the next six years we find him eking out some sort of an existence by casual coaching in Berlin.

Ultimately the first recognition of his researches came from abroad. Scientists in Russia, America and England began to discuss his papers, and presumably taking their cue therefrom, Germany at last brought Ohm out of his obscurity. In 1833 he was appointed to a lectureship at Nuremberg. The tide was now beginning to turn rapidly in his favour, and in 1841 he was honoured

by the Royal Society of London with the award of the
Copley Medal. Finally, in 1849, he achieved the crowning ambition of his life when he was appointed to a
professorship at the University of Munich, and he died
full of honours in 1854.

He left as a permanent tribute to his memory what
is universally known as Ohm's Law. It was the result
of theoretical deductions based on an analogy between
the flow of heat and the flow of electricity. Just as
in the former case there is, when heat is transmitted
through a slab of material, a falling off of temperature
as between the hot face from which the heat is sent,
to the cooler face receiving it, so there should be also
a falling off of some corresponding quality in the transmission of an electric current along a wire from one
terminal of the battery to the other. This something
is the fall in electrical potential. Ohm was thus led to
experiment on the conducting power of a wire. Now
both in the mathematical and the physical senses, the
reciprocal of the conducting power of the wire is its
resistance to the passage of a current. Ohm found
that the resistance offered to the passage of a current
is governed by its length (l), its cross-sectional area (a),
and the nature of the material. If R is the resistance,

Ohm found in effect that $R = \dfrac{\rho l}{a}$, where ρ was a constant

whose value differed from material to material. The
factors affecting the strength of current in any circuit
were then this *resistance* of the wire (and also of the
battery itself) and the difference in electrical potential
between the two battery terminals. Ohm spoke of the
' motive ' force due to this potential difference as the
' electroscopic force,' and it is nowadays referred to as
the electromotive force. A series of experiments brought
out a simple relationship between these three factors
in the electric circuit as follows :

$$\text{Current (C)} = \frac{\text{Electromotive force (E)}}{\text{Total Resistance (R)}}.$$

This is Ohm's Law, and it forms the basis of all calculations regarding electric circuits. It at once made possible the setting up and defining of units of current strength, electrical resistance, and of electromotive force, and its formulation made the world of physics a perpetual debtor to the genius of Ohm.

III. THE FOUNDERS OF ELECTRO-MAGNETISM

Thus far the phenomena of magnetism and of electricity had been the subjects of separate investigation, and there was no reason to suppose that they were in any way related. In 1819, however, a chance observation by Oersted (1777–1851), a Danish philosopher, opened up a completely new avenue of research that led to a direct linking together of these two groups of phenomena, with results that, on the theoretical side, have completely revolutionized the whole science of physics, and on the practical side have brought incalculable benefits and comforts to mankind.

Hans Christian Oersted was born in 1777 at Rud-kjobing, a small island town off the Danish coast. His father was an apothecary, and Oersted was thus able, from an early age, to become familiar with chemicals. In spite of much financial hardship he was able to enter the University of Copenhagen, and here his success was such that he obtained his doctorate at the age of twenty-two. A steady career of lecturing and research followed, and his interest was specially engaged in the consequences of Volta's discoveries. In 1806 he was appointed to the chair of Physics at Copenhagen, and his speculations began to lead him to a feeling, probably little more than an intuition, that magnetism, like galvanism, might be some ' hidden form ' of static electricity.

It was during the winter of 1819 that Oersted's famous observation was made. During the course of a lecture, he was exhibiting the heating effects of Volta's pile on a slender wire. There happened to be a magnetic needle, horizontally mounted, on the lecture bench, and by chance this needle was moved towards the battery and parallel to the wire. Oersted was astonished to see that at the instant the voltaic circuit was closed, the needle swung round and set itself almost at right angles to its original direction. On breaking the circuit the needle returned to its original state. He now reversed the direction of the current, and found that the needle swung again, this time in the opposite direction. Here, then, in this single observation, was brought out a remarkable fact. The magnetic behaviour of a needle was definitely influenced by an electric current. Oersted announced his discovery in a short pamphlet published in Latin in July, 1820. He was rewarded with the highest of honours both in his native country and else-where.

The sequel to the discovery was not long in arriving. Indeed, Oersted's successor in the sequence of investigation, André Ampère, announced further developments within a few days of his first news of the details. André Marie Ampère, a native of Lyons in France, was born in 1775. He was the son of a retired merchant, and showed signs of remarkable precocity as a child. His attraction to mathematical studies was irrepressible, and at the age of twelve he was already delving into the abstruse writings of Euler and Bernouilli. Much about this time he suffered much mental anguish due to his father's death on the guillotine. It is scarcely surprising that the strain of this, accentuated by the intensity of his reading, produced a severe though temporary break-down. In due course he recovered, and again threw himself into his studies. His marriage in 1797 prompted him to seek employment, and after holding minor appointments for a year or two he succeeded in 1805 to a post at the Polytechnic School at Paris. Here he gradually improved his position, and in 1809 he received a professorship at the school, a post he held until his death in 1836.

On 11th September, 1820, Arago, a distinguished member of the French Academy of Science, read a paper describing Oersted's experiment. Ampère was present and was profoundly impressed. He at once began experimenting for himself, and seven days later was able to announce important developments. He took a magnetic needle horizontally suspended and at rest in a north and south direction, and brought a wire forming part of an electrical circuit *over* it and parallel to it. On closing the circuit the needle swung to the west. He now broke contact, and brought the wire *under* the needle. On again completing the circuit, the needle swung to the east. Ampère was thus led to formulate his well-known rule : suppose a man to be swimming in the direction along which the current flows, so that the current enters at his feet and flows towards the head, and suppose that he is facing the magnetic needle (i.e., suppose the needle is under him) then the north pole of the needle will be deflected to the left by the influence of the current.

Ampère now clearly realized two parallel facts. A current affects a magnet, and a magnet affects a magnet. Why, then, should not a current affect a current ? This he now proceeded to test by experiment. Two wires were arranged, each forming part of a separate electrical circuit, so as to be both parallel and free to move. As soon as the circuits were completed so as to send the currents in the same direction, the wires were seen to approach each other. He now reversed one of the currents, and found the wires to move away from each other. Thus it was shown that surrounding every electric current is a magnetic field, and Ampère was able to deduce that if C and C^1 are the respective current strengths of two wires whose distance apart is d, then the force of attraction or repulsion between them is $F = \dfrac{CC^1}{d^2}$. It was this inverse square law that prompted Clerk Maxwell to refer to Ampère as ' the Newton of electricity.'

From these researches there quickly emerged the galvanometer, a new means of measuring current strength

8

in terms of the force with which a magnetic needle situated at the centre of a coil of wire is deflected over a graduated scale when the current to be measured is made to pass through the coil, and it is a fitting honour to the memory of this great physicist that the practical unit of current has been named after him.

In the broad sequence of events that brought into being the basic facts of electro-magnetism, there was one final step awaiting accomplishment. Ampère had shown how magnetic phenomena were produced by electricity. It now remained to show, conversely, that electrical phenomena could be produced from magnetism. The honour of this great achievement belongs to one of the most romantic figures in the history of physics— Michael Faraday.

Michael Faraday was born at Newington, in Surrey, in 1791. His father was a poor working man, a blacksmith by trade, and when Michael was five years old the family moved into rooms over a coach-house in London. Here they lived a life of great poverty. Michael learnt the rudiments of reading and writing at a ' common day school ' until he was thirteen, when he was apprenticed to the bookbinding trade. It was this intimate contact with books that opened out for him visions of a larger world. He found himself reading a book on *Conversations in Chemistry*, and science claimed him then onwards. Minor experiments with home-made apparatus and the attending of popular science lectures served both to fan his hopes and ambitions and to increase his distaste for bookbinding. At the age of twenty-one he decided on a bold step. Sir Humphry Davy was at this time Director of the Royal Institution, and to him Faraday wrote, as he tells us, ' expressing my wishes, and a hope that, if an opportunity came in his way, he would favour my views ; at the same time I sent the notes I had taken of his lectures.' The appeal was successful, and in 1813 Faraday was appointed as Laboratory Assistant at the Royal Institution. Thus was brought into the world of science a man destined to bring it honour.

We have not the space, unfortunately, in these pages, to recount step by step the details of his steady advancement. Suffice it to say that from the washing of bottles and the care of apparatus he rose steadily by sheer worth and merit, and ultimately in 1845, two years after his election as a Fellow of the Royal Society, succeeded his great patron, Sir Humphry Davy, as Director of the Royal Institution. Devoid of mathematical ability, he yet became a recognized prince of experimental science, and although his discoveries have been proved to possess immense commercial possibilities, he died a poor man. He scorned to seek either financial reward or scientific favours. The former he consistently refused. Of the latter many were showered upon him in spite of himself. 'I must remain plain Michael Faraday to the last,' he said. In 1862 he retired from the Royal Institution to apartments at Hampton Court

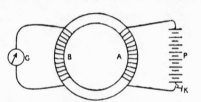

Fig. 20.—Faraday's Experiments in Electro-magnetism

placed at his disposal by Queen Victoria, and here he died peacefully in his seventy-sixth year in 1867.

What was the nature of Faraday's monumental contributions to the science of electro-magnetism? With a full knowledge of the experiments of Oersted and Ampère, he had for some time felt that it ought to be possible to produce the converse effect of deriving an electric current from magnetism. These views occupied his attention but intermittently from 1824 to 1831. It was in this latter year that success at last came to him. Ampère had shown, among other things, that a cylindrical coil of wire, known as a solenoid, behaves exactly as if it were a magnet when a current is passed through it, and Faraday decided to utilize this fact. He took an

iron ring $\frac{7}{8}$ inch thick and six inches in diameter (Fig. 20)
and at each half of it he wound separate coils, A and B,
each of about 70 feet of copper wire. A was connected
up with a voltaic pile P, and B with a current detector
or galvanometer G. The two circuits were thus completely separate. A key K made it possible to make or
break the voltaic pile circuit at will. On depressing the
key, a current flowed round A, and the galvanometer
connected to B at once began to swing round, and then
return to rest in its original position. At the instant
K was released, thus breaking the circuit, the needle
of G again swung round, this time in the opposite direction from before, after which it again came to rest.

Here then was a remarkable achievement. By passing
a current round
the solenoid B,
Ampère had
shown that it,
and therefore
the whole iron
ring, behaves
like a magnet;
and it was this
magnet that de-

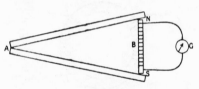

FIG. 21.—Faraday's Experiments in
Electro-magnetism

finitely induced a momentary current of electricity
round circuit A, and it was this current that was registered by the galvanometer.

Faraday now decided he must try to achieve his
results without the aid of a battery at all. He succeeded
brilliantly within a month or two of his first experiment.
He wound a coil of wire round a short bar of soft iron
and thence on to the terminals of a galvanometer G
(Fig. 21). He then placed this between two bar magnets
A N and A S, so that opposite poles were at each end
of the bar. The magnets were in contact at A. *Every
time the contact between the iron and the magnets was
made or broken the needle in G indicated the momentary
passage of a current*, the direction of the current at
'break' being again opposite to that at 'make.' Here,

definitely, was the converse effect to Oersted's that he had been seeking.

He then proceeded to his third experiment, his object being to dispense with any source of magnetism whatever, whether from a solenoid connected to a battery, or from a magnet direct, as had been the case previously. This time he substituted for the iron ring of Fig. 20 a wooden bobbin. Round this he wound a coil of 203 feet of copper wire, and connected it up to a voltaic cell. Round this ' primary ' coil was wound a second and similar coil (the ' secondary '), its ends being joined to a galvanometer. As before both on ' make ' and ' break ' momentary currents were registered through the galvanometer. On the ' make ' a current was sent round the inner coil, and this set up a magnetic field round it, as was shown by Ampère. It was the influence of this magnetic field on the outer coil that induced the momentary electrical effects. Again, too, the effects were opposite at ' make ' and ' break.' It was clear, furthermore, that these induced effects were always associated with a *relative change of conditions*, e.g., at the ' making ' of the circuit, and at the ' breaking ' of the circuit, and Faraday explained that the current is always induced in a wire wherever the ' lines ' of a magnetic field of force ' cut ' the wire. Thus in the third experiment, on the break, the current is started in the primary, and it therefore ' throws out ' its magnetic field, and as soon as this field ' cuts ' the secondary ' outwards ' the induced current is set up. On the ' break ' the current ceases, the magnetic field withdraws ' inwards,' and in so doing ' cuts ' the secondary in the opposite direction to the previous one, and so the induced current is set up in the reverse direction.

This then was Faraday's great work ; and this slight and apparently insignificant kick of the galvanometer needle was the starting point of the magneto, the dynamo, the induction coil, the electric motor, the telephone, and a whole host of important applications which have so greatly altered and augmented the amenities of modern

life. Faraday's own vision of the possibilities of the
' kick ' must have been great, for to Gladstone's bored
question as to what use it could be, he answered, ' Sir,
you will soon be able to tax it.'

We have not attempted to recount more than a fraction
of this great physicist's contributions to modern science.
We have merely confined ourselves to one aspect in the
development of the theme of this chapter. At least,
however, enough has been said to make it obvious that,
in the words of Tyndal, ' his was the glory of holding
aloft among the nations the scientific name of England
for a period of forty years.'

Vast as Young's papers were in importance, they formed the subject of a violent and ill-founded attack by Lord Brougham in the *Edinburgh Review*, and in spite of an able and a complete reply by Young, public opinion was definitely biased against his theories.

II. STEPHEN LOUIS MALUS (1775–1812)

Unfortunately, too, a new discovery by Malus, a Frenchman, regarding light was now at hand which seemed to offer further difficulties to the undulatory theory. Stephen Louis Malus was a French mathematician who was educated at the famous *Ecole Polytechnique* at Paris. He entered the French Engineering Corps at twenty-one and saw much active service, experiencing many campaigning hardships in Egypt and Syria. From 1802 to 1807 he was engaged in constructing fortifications at Antwerp and Strasbourg. He died at the early age of thirty-seven in 1812.

The discovery with which his name is associated is that of the *polarization of light by reflection*. One day in 1805 Malus was viewing the rays of the sun reflected from the windows of the Luxembourg Palace through a doubly-refracting crystal of Iceland spar. He expected, of course, to see two images. In fact he saw one only. There was, however, this interesting point, that in one position of the crystal he saw what was obviously the ' ordinary ' image, but in a position at right angles to this he saw the extraordinary image. Between these two positions he got two images, one much brighter than the other. That evening he tested matters further. He found that the light of a candle reflected from a glass plate was, at an angle of about 35°, as completely ' polarized ' (as he termed it) as had been the rays emerging from the Iceland spar ; from water he found the rays polarized at an angle of about 36°, and generally that light reflected from all *transparent* bodies is always more or less polarized.

As a matter of fact Huyghens had, one hundred years

before, observed similar effects when passing the ' double'
rays from an Iceland spar through another crystal
after Bartholinus first discovered the phenomenon of
double refraction, and Huyghens suggested, correctly
as we now know, that the crystal was more elastic in
one direction than in the other, thus causing one wave
passing into a crystal of Iceland spar to be divided into
two waves moving at different speeds through the
crystal. Huyghens' observations were by now forgotten,
however, and the whole matter called for fresh explanations, in the light of the fact that polarization of light
was possible by reflection as well as by the action of
crystals.

Young found himself unable to explain the phenomenon by his theory, and in 1811 wrote to Malus to
this effect, adding his view that it nevertheless did not
prove the falsity of his theory. We shall now see that
both this and other aspects of the undulatory theory
finally received the fullest verification at the hands of
another illustrious Frenchman, Auguste Jean Fresnel.

III. AUGUSTE JEAN FRESNEL (1782-1827)

Fresnel was born at Broglie in Normandy in 1788.
By way of contrast with Young, he could scarcely read
at the age of eight, and as a child was always of delicate
health. He was educated first at a school at Caen,
and afterwards at the *Ecole Polytechnique*. He decided
on the profession of civil engineering in the service of
the State, and after a course at the *Ecole des Ponts et
Chaussées*, he served the Government in the repair
and construction of roads over the recently devastated
area of La Vendée. He found the work none too
pleasant. 'I found nothing so laborious,' he writes,
' as leading men, and I confess I know nothing at all
about it.' It was by way of relief from the tedium of
his employment that he began to study the theory of
light. In 1814 he joined the small army that attempted
to oppose the landing of Napoleon. He was rewarded

by being deprived of his post. This gave him leisure
for more study and research, and we find him sending
for books ' on the polarization of light,' about which
he confessed he knew nothing. Eight months later
his researches in this very field had made him famous !
He developed a positive genius for experiment. He
made a micrometer with threads and pieces of card-
board, and with such rough but accurate apparatus
he was able to present two memoirs to the Academy
of Sciences at Paris. These brought him his reinstate-
ment in Government employment in Paris, and led to

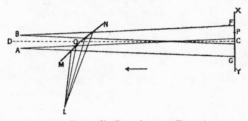

FIG. 23.—Fresnel's Interference Experiment

his appointment as Secretary to the Lighthouses Commission. In 1819 he gained the Academy prize for
an essay on diffraction, and in 1823 he was elected a
Member of that body. The Royal Society of London
also honoured him—with a foreign associateship in 1825
and the Rumford Gold Medal in 1827, and a few days
after receiving this last distinction he passed away,
at the early age of forty-five.

To Fresnel belongs the honour of having set the
undulatory theory on the sure basis of quantitative
determinations. His experiments led to the formulation
of the laws of the phenomena mathematically. His
most famous experiment was with the employment
of inclined mirrors, and it supplies the most direct
method of studying the interference of light. In Fig. 23,
O M and O N are the two mirrors, almost in a straight

line. At L rays of sunlight enter a dark room through a very narrow slit, and fall on the mirrors. They are thence reflected on a white screen F G. Looking into the mirrors from the screen, two virtual images A and B of L are seen in accordance with the ordinary laws of reflection, and the effect is just the same as if these are the two light sources for the production of interference effects. By a slight rotation of one of the mirrors, the two light sources (i.e., the two images) may be brought as close together as required. Consider a point P on the screen receiving rays from both sources. Clearly the distance A P is greater than B P, and (A P — B P) becomes less the nearer P is to C. It is a matter of simple mathematics to calculate the distances B P and A P, since P C, O C, and O L and the angle that A B subtends at C are all directly measureable. Hence the difference between the two paths A P and B P is known, and in the case of monochromatic light, i.e., light of one colour, as in previous examples, P will show a dark band or a light band according as (A P — B P) is an odd or an even multiple of half-wave lengths; a bright band is always at C. Hence the interval between two successive bright or dark bands depends upon the value of the wave-length. Therefore, by measuring this interval between two consecutive bright or dark bands, the wave-length of the particular colour of light employed may be calculated. In this way Fresnel was able to discover the value of the wave-length of waves corresponding to all the colours of the spectrum—a triumph for the wave theory, because, among other reasons, it met the objection of opponents who insisted on explaining diffraction phenomena on the emission theory by producing two point-sources of light without the use of apertures or strips of opaque objects; in other words, by eliminating diffraction effects completely.

It is refreshing to note that only after this work did Fresnel become acquainted with Young's prior researches, and yet, so far from there being any jealousy between the two, they became firm friends, and maintained a close correspondence.

What we may speak of as the mechanism of the wave theory now called for closer attention, especially having regard to the necessity for explaining the phenomenon of polarization. Collaborating with Arago, another famous physicist, Fresnel found that the ordinary and the extraordinary rays emerging from Iceland spar would *not* interfere with each other. This at once led him to suspect that the waves in the rays must move in a different manner. Discussion with Arago and correspondence with Young brought out the important conception of *transverse* vibrations. In a sound wave the vibrating particles are moving to and fro in a direction parallel to the propagation of the wave. This is known as longitudinal vibration. In a water wave the water particles are moving up and down at *right angles* to the forward direction of the wave. This is known as *transverse vibration*. The

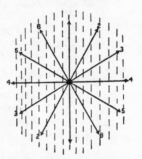

Fig. 24.—To illustrate the meaning of Transverse Vibrations

ether vibrations of a light wave are of this latter type, but with this important complication, that the plane of vibration is not restricted, so that a ray of light may consist of waves vibrating in every conceivable plane at right angles to the direction of propaganda. Fig. 24 will help to make this evident. Looking ' end-on ' behind a single wave, we merely see a short straight line showing the extremities between which the ' ether particle ' vibrates. The figure shows a group of six such light waves. Although they are vibrating in planes at all angles, yet each vibration is at right angles to the direction in which the wave is advancing, i.e., forward into the plane of the paper.

We can now understand the action of Iceland spar, which acts upon the light impinging on it as though a set of railings (shown dotted in the figure) were interposed. The *one* ray vibrating parallel to the rails will pass on, and the remainder will be stopped ; and the ray that passes on is the plane-polarized ray. In the case of Iceland spar, of course, there are *two* rays that pass through, vibrating at right angles to each other.

Fresnel and Arago were able to show, by experiments and demonstrations that are beyond the scope of this book, how such polarization results from reflection on the basis of the undulatory theory.

The final deathblow to the emission theory, practically abandoned by 1825, was given on 6th May, 1850, when Foucault was able to determine accurately the relative velocities of light in air and in water. It will be recalled that the emission theory required that the velocity in water must be greater than in air, whilst the undulatory theory required the converse to be true. Foucault definitely found the velocity in water to be less than in air, and so the undulatory theory triumphed.

CHAPTER XI

THE DOCTRINE OF ENERGY

I. CALORIC

WE have referred to the ' doctrine of imponderables ' which became so general a feature of science in the eighteenth century, and we have mentioned that heat was one of the phenomena that was brought within the scope of its conceptions by the theory of ' caloric.' Let us review briefly the main outlines of this theory. ' Caloric ' was held to be an imponderable fluid contained within the pores of all substances, and this idea persisted firmly as an accepted doctrine for just as long as it continued satisfactorily to explain all observed facts with regard to heat. We shall see in this chapter that the time did at last come when certain new facts emerged which could not be satisfactorily explained by the ideas of ' caloric,' and as a result it had to share the fate of the other members of the family of imponderables in being discarded. ' Caloric,' it was assumed, was squeezed out of bodies on heating. Black's researches on calorimetry were conducted in full acceptance of this theory, and indeed it fitted the facts. When a hot body is mixed with a cold liquid, for example, ' caloric ' passes from the hot to the cold, cooling the one and heating the other, until the two are at one common temperature. Friction, again, had long been known to be accompanied by the production of heat. The explanation was simple. If a metal is hammered, the ' caloric ' is forced out of its pores, and the metal gets hot. If a piece of metal is

bored, the chippings, being small, cannot contain so
much caloric as when forming part of the main piece,
and they get hot. Even Black's experiments on latent
heat, at first apparently difficult to reconcile with the
doctrine of 'caloric,' were made to 'toe the line.'
The explanation here was that the caloric combines
with the ice to form water, which was therefore regarded
as being compounded of ice and caloric. Yet another
view of the nature of heat had been making sporadic
appearances—the view that heat is matter in motion.
Democrtus and Epicurus had speculated on this in
remote Greek times. We have also seen that Francis
Bacon concluded much the same thing. So in fact
had Boyle, Gassendi, Newton and Huyghens. This
was the view that was, in fact, destined to prevail,
and the first great physicist to help in this achievement
was Benjamin Thompson, Count Rumford.

II. COUNT RUMFORD (1753–1814)

Benjamin Thompson was born in 1753 at North
Woburn, a small village in Massachusetts—within a
few miles of the birthplace of his famous fellow-country-
man, Benjamin Franklin. His early life was full of
variety. From thirteen to eighteen he was an assistant
in a dry-goods store, during which time he began to
develop a taste for study and experiment. He then
became successively a medical student (Harvard College,
Cambridge, was but eight miles from North Woburn)
and a schoolmaster. At twenty he had married a
wealthy widow, from whom, however, he shortly afterwards separated. Then the War of Independence
broke out. Although he was a major in a militia
regiment, he was suspected of Tory leanings and
'unfriendliness to the cause of liberty.' In November,
1774, he was nearly mobbed at his house, but had fled
to Boston, where he was arrested and charged. The
charge failed, and he was released. In the autumn of
1775 he was sent to England with despatches from

General Howe to the British Government, and now parted company with both his wife and America for ever. In fact he now participated in the war in an under-secretarial capacity on the side of the British. In London Thompson conducted experiments with gunpowder and interested himself in other scientific topics. As a result he was elected a Fellow of the Royal Society. Military life apparently held a fascination for him, however, and in 1783, while travelling on the Continent, he was invited at Munich to enter the service of the Duke of Bavaria. With the permission of King George III (who knighted him for his services) he accepted, and in 1791 the Duke of Bavaria raised him to the dignity of Count of the Holy Roman Empire as Count Rumford. His career was now extraordinary. He became Head of the War Department, was made a State Councillor, controlled almost every branch of the public services of Bavaria, founded schools of industry, a military academy at Munich, improved the lot of the soldier, reduced beggary, and all the time continued his scientific studies.

In 1798, after an absence of eleven years, he was sent to England as Minister Plenipotentiary of the Court of Bavaria. As a British subject, however, this appointment could not be ratified, and Rumford decided to remain in England for a few years in a private capacity. Here he took a prominent part in founding the famous Royal Institution in January, 1800.

In 1803 Rumford went to France, and two years later married Madame Lavoisier, widow of the famous chemist. It was an unhappy union, and a divorce followed four years later. Count Rumford died at Auteuil, near Paris, in 1814.

In 1798 Rumford wrote a paper on ' An Inquiry Concerning the Source of the Heat which is excited by Friction,' in which he describes how, while engaged in superintending the boring of cannon at Munich, his attention was arrested by the great amount of heat generated in the process. He proceeded to experiment.

Thus when a brass cylinder was made to revolve against a steel borer, the cylinder being placed inside a wooden box containing 18¾ lb. of water so that the amount of heat absorbed could be registered, and at the same time preventing any of the heat from being lost, the temperature rose in 2¾ hours from 60° F. to boiling point. ' It would be difficult to describe,' said Rumford, ' the surprise and astonishment expressed in the countenances of the bystanders on seeing so large a quantity of water heated, and actually made to boil, without any fire.' Whence, he asked, came this heat produced by a purely mechanical operation ? ' As the machinery used in this experiment could easily be carried round by the force of one horse . . . these computations show further how large a quantity of heat might be produced, by proper mechanical contrivance merely by the strength of the horse, without fire, light, combustion or chemical decomposition.' The fact that his apparatus was immersed in water proved that the heat did not come from the air. Nor could it, he argued, be derived from any diminished capacity for heat from the chippings, since on collecting and testing them he found no sign of any change in their capacity for heat. Again, the store of ' caloric ' in the brass cannon should surely ultimately be reduced to nil if the ' caloric ' doctrine were right ; yet in fact he found no limit to the production of heat.

Rumford came to the inevitable conclusion that the ' doctrine of caloric ' must be wrong, and recalling the other theory that had occasionally been offered, he concluded definitely as follows :

' It appears to me extremely difficult, if not quite impossible, to form any distinct idea of anything capable of being excited and communicated in the manner the heat was excited and communicated in these experiments, *except it be motion.*'

Rumford's researches were strikingly amplified a year later by Sir Humphry Davy (1778–1829), famous alike as physicist and chemist. Davy arranged an apparatus by means of which two pieces of ice could be made to rub against each other in the exhausted receiver of an air-pump. In a few minutes the pieces of ice were entirely converted into water which, when collected, was found to be at a temperature of 35° F., although the temperature of the atmosphere was lower than this. Now according to prevalent notions, the capacity for heat had diminished. Davy pointed out, however, that the capacity for heat of water is about *twice* that of ice, and ice must have an absolute quantity of heat added to it before it can be converted into water. Hence friction could not diminish the capacity of bodies for heat. Further, the heat in this experiment could not have come from the air, since the experiment was performed *in vacuo ;* nor could it be derived from the apparatus, since the water formed *only* at the ice-surfaces in contact. By 1812 Davy had become sufficiently convinced to conclude that ' the immediate cause of the phenomenon of heat, then, is motion, and the laws of its communication are precisely the same as the laws of the communication of motion.' Some thirty years were to elapse, however, before the next step in the construction of the new doctrine of energy.

III. SADI CARNOT (1796–1832)

It is curious that the next contribution to the subject of our study in this chapter should be made by an adherent, as he was at the time, of the caloric theory. Such was, however, the case with Nicolas Léonard Sadi Carnot, a young French engineer. By now the steam engine had been invented, and Carnot was naturally interested in the scientific problems raised by the invention. He was fully aware of the fact that mechanical work could be produced by heat, and in seeking for an explanation of this he was impressed with an analogy between the behaviour of heat in relation to the production of mechanical work and that of the mechanical power of a head of water. He argued that just as work is done by the latter *only* in its descent

from a higher to a lower level, so work done by heat
is necessarily attended by a fall of some body from a
higher to a lower temperature. Consequently, he said,
whatever might be the nature of the substance of the
body, whether it be steam, air, vapour, gas or liquid,
this fall in temperature is essential to the production
of work, and the amount of work done depends on this
fall and is independent of the nature of the substance
to be employed. For example, just as water can be
made to perform useful work by taking it in at a higher
level, and discharging it, say, by an overshot wheel,
to a lower level, so we can obtain work by taking in
heat at a high temperature in, say, the boiler of a steam-
engine and discharging it at a lower temperature to
the condenser. There is here, so far as Carnot is concerned, no loss of heat. Caloric is merely transferred
from boiler to condenser ; but the important point
—and this is independent of any theory of heat—is
that the amount of work which can be done depends
solely on the quantity of heat passed through the
engine and the temperatures of the boiler and condenser.

Carnot published his views in 1824 in a work called
Reflexions sur la puissance motrice du feu. In this
work he developed the important idea of a *reversible
cycle of changes* as a scheme for determining mathematically how much mechanical work could be performed not alone by a steam engine, but more generally,
by an ideal ' heat-engine.' The cycle of operations
consists in starting with the heat-source—the boiler—
at the high temperature, extracting heat from it, then
utilizing *part* of this heat for conversion into mechanical
work, and passing the remainder to the condenser at
the lower temperature. Then, reversing the cycle,
he performs *on* the engine (i.e., by driving the engine)
an amount of work equal to that previously performed
by the engine, and in so doing is able to extract the
heat again from the condenser, and restore it to the
original heat-source. Theoretically, in this ideal engine,
steam need not necessarily be employed. In fact any

substance which is a source of heat, and which will enable the passage of some of this heat from it to some ' receiver ' at a lower temperature, will do equally well. What clearly emerged, among other important consequences, from Carnot's important researches (and unfortunately they received little attention from the scientific world for a long time), was that to a given amount of mechanical work there corresponds a definite amount of heat. Carnot, however, did not himself measure the value of this ' equivalence.' The honour of this achievement belongs instead to an English scientist, James Prescott Joule, whose life and work therefore next claim our attention.

IV. JAMES PRESCOTT JOULE (1818–89)

Joule was born in Salford in 1818. He was the son of a wealthy brewer, and being of somewhat delicate health as a child, he was educated at home until he reached the age of sixteen. Together with his brother, he then continued his studies under the famous chemist, John Dalton, and under the inspiration of this great teacher he not only studied mathematics, but developed a taste for original research. Accordingly his father provided a laboratory for him in his house, and thence onwards he devoted all his spare time to the cause of science. His first results were concerned with an effort to improvise an electro-magnetic engine, and from this he was led to the measurement of the electric power absorbed by such an engine. This line of research led to important results, and towards the end of 1840 he produced a paper ' On the Production of Heat by Voltaic Electricity,' dealing with the heating effects of a wire carrying a current. He showed that the heat generated is proportional, in any given time, both to the resistance and to the square of the current.

Joule was now beginning to dwell upon problems of heat in relation to mechanical energy. In 1843 he wrote a paper ' On the Mechanical Value of Heat ' in which he remarks :

' Having proved that heat is generated by the magneto-electrical machine, and that by means of the inductive power of magnetism we can diminish or increase at pleasure the heat due to chemical changes, it became an object of great interest to inquire whether a constant ratio existed between it and the mechanical power gained or lost.'

An electro-magnet contained in a tube filled with water was rotated between the poles of a powerful magnet, and the increase in temperature was noted. From this it was possible to calculate the number of heat units absorbed by the water from the electro-magnet. In order next to find the amount of mechanical work done, it was necessary to measure the force required to turn the apparatus. This was found by winding twine round the spindle carrying the electro-magnet, then passing the twine over a pulley and attaching a scale pan to the end of it. The difference between the weights required to produce equal velocities, with the induced circuit complete and then interrupted, gave the force. Thence the work done against the electrical forces, which reappeared as heat, could be calculated. The experiment was difficult and led to many sources of error. Joule found in an average of thirteen tests that ' the quantity of heat capable of increasing the temperature of a pound of water by one degree Fahrenheit is equal to, and may be converted into, a mechanical force capable of raising 838 lb. to the perpendicular height of one foot.' Nevertheless he recognized the need for a more accurate determination, and in 1845 he measured the work done and the heat generated when air is compressed—an experiment previously performed in Germany by Mayer—and he examined the converse experiment by Seguin of allowing air to expand.

In 1845 he also measured the heat produced by the agitation of water when a paddle wheel revolves in it, and in 1847 he finally adopted this method on most refined lines as the best available. This is illustrated in Fig. 25. A is a vessel which may be filled with water,

oil, or other liquid. The spindle *s* passes through the
cover of the vessel and carries four vanes at right angles
to each other, cut as shown by the dotted lines. These
vanes pass through corresponding apertures in eight
fixed radial partitions, the object of which is to prevent
the liquid from whirling round with the rotating paddles.
The spindle *s* carries in its upper part a cylinder *c*
which turns with or without the spindle as the peg *p*
is inserted or withdrawn. The cords $t\,t^1$ attached to
the drums $d\,d^1$ are wound upon the cylinder, and the
descent of the weights $w\,w^1$ therefore causes the spindle
to rotate. The
heights through
which the weights
descend are meas-
ured by the scales
r r^1, and the
weights are then
again raised into
position with the
peg *p* withdrawn,
and the experi-
ment repeated a
number of times.
Joule's final value
was 772 foot-lb.

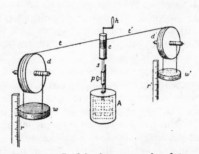

Fig. 25.—Joule's Apparatus for determining the Mechanical Equivalent of
Heat

of work as the *mechanical equivalent* of heat which
will raise the temperature of 1 lb. of water one degree
Fahrenheit. The modern value is 778.

Joule was elected a Fellow of the Royal Society in
1850, and was President of the British Association in
1872. He died in 1889, and a memorial tablet has been
erected in his honour in Westminster Abbey.

V. LORD KELVIN (1824–1907)

The final step in the story of the doctrine of energy
brings us to one of the greatest figures in nineteenth-
century physics, William Thomson, Lord Kelvin.

Thomson was born in 1824 at Belfast, his father being a Professor of Mathematics at the Royal Academical Institute in that city. In 1832 Professor Thomson was offered a professional appointment at the University of Glasgow, and thither the family accordingly migrated. Young Thomson proved himself an ardent student, and matriculated at the age of ten. In 1841 he went to Cambridge, and after a brilliant academic career came out second in the Mathematical Tripos lists, and shortly afterwards won the Smith's Prize —the highest mathematical award of the University. In 1846, at the early age of twenty-two, he was appointed Professor of Natural Philosophy at Glasgow University, and achieved what must surely be a record in academic history by holding the one appointment for over sixty years.

The range and importance of his physical researches were remarkable, and space does not permit even of their enumeration. He was also brilliant—and incidentally business-like—in the field of applied science. To him belongs the honour, for which he was knighted, of having laid the first Atlantic cable. Navigation was another of the branches of applied physics which benefited enormously by his activities. He designed a depth-sounding machine that earned for him the gratitude of all seamen in enabling many errors in charting to be rectified. In one instance when a ship was, according to the chart, in water showing 'no bottom' at 100 fathoms, the captain gave the order, 'Throw Thomson overboard!' and, as he relates the incident, 'I heard the whirr of the wire suddenly stop. Fifteen fathoms! . . . I hove to all right—but had we continued on our course we should have been ashore before daylight. The coast line was not correctly charted. No wonder we sailors bless the name of Lord Kelvin.'

Kelvin was elected a Fellow of the Royal Society in 1851, and became its President in 1890. Honours were bestowed on him from almost every country in the world, and in 1892 he was created a peer as Baron Kelvin of Netherall, Largs. The high esteem in which he was held by the whole world was evidenced by a remarkable demonstration at Glasgow in honour of the fiftieth year of his professorship, at which practically every country and scientific society was represented. In 1899 Kelvin, at the age of seventy-five, resigned his appointment, but kept his name on the rolls of the University as a research student. He died on the 18th December, 1907, and very fittingly was buried at Westminster Abbey by the side of the grave of Sir Isaac Newton.

When Joule read his now famous paper at the Oxford meeting of the British Association in 1847, it threatened to pass unnoticed. It was William Thomson who brought it the publicity it merited by recognizing that it constituted a great event in the history of science. Kelvin saw a discrepancy between Carnot's theory and Joule's results, and it was through his insight that both were seen to fit in with each other to produce, in their aggregate, the two laws of thermodynamics which form the basic facts underlying the doctrine of energy.

Carnot had written his important essay on the assumption of the caloric theory (it should be mentioned that he abandoned this theory before he died), and although his main thesis was independent of any theory of heat whatsoever, there was at least one assumption contained in it which offered a serious difficulty. According to his view, no loss of heat was involved in the transfer of caloric from the ' boiler ' of his ideal engine to the ' condenser.' Yet the engine is made capable of performing work by the passage of a quantity of heat passed through it. Now heat passes continually from the fire in a room, say, to the cooler walls and surroundings. Where, then, is the work that it should have done ? Here was an incongruity in connection with which Kelvin first used the term *energy*. ' Nothing can be lost in the operations of Nature—no *energy* can be destroyed.' It was Kelvin, who, quick to seize on the significance of Joule's work, saw that in Carnot's cycle there *was* definitely a portion of the heat destroyed as such. It was a quantity equal to the difference between the amount of heat taken from the boiler and the amount of heat received by the condenser, and this was exactly equal, in terms of Joule's value of the 'mechanical equivalent of heat,' to the amount of mechanical work done by the engine.

Joule's law, that there is a definite numerical equivalence between heat and work, led Kelvin and others to formulate the great generalization in physics known as the *Law of Conservation of Energy.* The sum total of the energy of the universe is constant and unalterable, so that whenever heat disappears as such, the loss of energy this represents is at once counterbalanced by an equal gain in energy in some other form. Mechanical energy is merely one of many such forms. Others are chemical energy, radiant energy, and so on. Our everyday life teems with illustrations of the conversions of forms of energy brought into play by our engineers for the benefit of mankind. The accumulated chemical energy of countless ages which we find in the form of coal is converted by combustion into heat energy. This is in its turn utilized in the steam engine for conversion into mechanical energy, and possibly this mechanical energy is made to drive a dynamo to produce electrical energy. Part of this electrical energy may be carried to 'electric motors' underneath tramcars for conversion into mechanical energy, and part is transmitted to our houses for change into light energy for illumination—and so on.

There was, however, one final qualification of the law of the conservation of energy to which Kelvin drew attention. The truism which he formulated as the second law of thermodynamics, 'it is impossible by means of inanimate material agency to derive mechanical work from a body by cooling it below the temperature of the coldest part of its surroundings' carries with

it the truism that heat is in fact the least useful of all
forms of energy. Every time a transformation of
energy is attempted, *some* of it is converted into heat
in a form which renders it virtually lost to utility.
If we carry a current along a wire, the current gets hot,
and the electrical energy received at one end of the
circuit is less than that sent from the other end by an
amount equivalent to that of the heat generated in the
wires. In every mechanism employed, the friction of
the parts generates heat which is lost at the expense
of the energy put into the mechanism. Every form of
energy known pays its toll in this way, the heat so
generated being conducted or radiated away into space.
So much is this the case that of the coal used in an
ordinary house fire, only about 7 per cent. of it is usefully
utilized for the purpose for which it is intended. Perhaps
the vast stores of solar energy radiated out into the
universe from the sun is as striking an illustration as
any. Only a minute fraction of it is received by the
earth. The remainder is lost to mankind, dissipated
out in space. So we get Kelvin's famous Principle of
the Degradation of Energy, which tells us that though
the total quantity of energy in the universe is a constant,
yet it persistently becomes less and less available for
useful work by virtue of its steady distribution throughout the universe in the form of heat. There is something startlingly dramatic in the ultimate fate of the
universe to which Kelvin's researches thus bring us—
that of a stagnant world of stillness in which all the
energy will have been evenly and uselessly distributed
throughout space.

VI. CONCLUSION

Limitations of space compel us to a close. It must
not be thought, however, that we are at the end of the
story of physics. The narrative we have offered is in
no sense a history of physics—nor is it intended as such.
We have, in fact, merely confined ourselves to certain

broad movements in the development of physical science, and in so doing we have referred only to a fraction of that great army of the physicists of the past whose memory it has been our purpose to honour. We have discussed but a chosen few of the large number of discoveries which these true benefactors of mankind have contributed towards the ultimate knowledge of the facts of life and the universe. That the search is well worth while is a commonplace of civilization. That the details of its seeking are all too little known is, perhaps, a sad commentary upon certain aspects of what is understood by the term ' a liberal education.'

Meanwhile, uninterrupted by the comings and goings of everyday life, the alarums and excursions of national and international relationships, the work of research goes on. Undeterred by the many disappointments inevitable in the progress of investigation, and always with a goal in view that he knows to be worthy of endeavour for its own sake alone, the man of science pursues his self-allotted task ; and, inspired by the examples of the many great seekers of the past, he pursues it faithfully and unfalteringly to the end.

The establishment of the doctrine of energy was the outstanding contribution to knowledge made by the physicists in the nineteenth century. It completely revolutionized the whole of the outlook on science, and it at once gave to what had hitherto been regarded as a number of detached phenomena a relationship that brought a new and vivid light to our understanding of nature. Yet such is the quickening pace of research to-day that all this has been left far behind. Modified in form, the doctrine of energy remains, but the knowledge of its machinery is incomparably more thorough. It has brought in its train a host of further problems of funda-mental importance, together with a large number of practical applications and devices for the increased comfort of mankind. It has brought about, too, a curious and interesting return to earlier conditions. We have seen that in its early days physics was not known as such. It belonged, in common with all problems of science, to the general field of natural philosophy.

As knowledge increased, this all-embracing natural philosophy separated out into its various branches of science—chemistry, astronomy, mechanics, heat, light, sound, magnetism and electricity and so on. These 'subjects' pursued their separate paths, and were 'adopted' by their separate investigators. The consequences of the doctrine of energy have once more united these studies into a common whole. Researches on radiant energy have shown how much there is in common between heat, light, magnetism and electricity; the discovery of radium and the wonderful researches on the electronic structure of matter have linked up physics and chemistry in their absolute fundamentals. The physical picture of the atom itself, with its elaborate and beautiful system of a central nucleus whose attendant electrons move round it in closed orbits, brings it into immediate comparison with the giant stellar systems of astronomy—our own solar system, with its central sun and rotating planets is but one of hosts —and shows it to be truly a wonderful instance of a universe in miniature. Nor must we forget that the final link of unity in all this is the machinery of mechanics —the laws of motion, obeyed alike by giant suns and minute electrons. Here, too, the pace of research has quickened. Our whole conception of the relations and meaning of time and space have but recently undergone radical change through the researches of Einstein and others, the final consequences of which are by no means yet in sight.

So for the present we may leave our review of the physical sciences, with feelings of reverence for those men of the past who have brought so much of the light of knowledge to the windows of our understanding, and of respect for the physicists of to-day and to-morrow, giving what encouragement lies in our power to the work that will bring fresh light and still fresh light to those who will live after us.

INDEX OF CHIEF NAMES